NONIONIC SURFACTANTS

SURFACTANT SCIENCE SERIES

1. Nonionic Surfactants, *edited by Martin J. Schick* (see also Volumes 19, 23, and 60)
2. Solvent Properties of Surfactant Solutions, *edited by Kozo Shinoda* (see Volume 55)
3. Surfactant Biodegradation, *R. D. Swisher* (see Volume 18)
4. Cationic Surfactants, *edited by Eric Jungermann* (see also Volumes 34, 37, and 53)
5. Detergency: Theory and Test Methods (in three parts), *edited by W. G. Cutler and R. C. Davis* (see also Volume 20)
6. Emulsions and Emulsion Technology (in three parts), *edited by Kenneth J. Lissant*
7. Anionic Surfactants (in two parts), *edited by Warner M. Linfield* (see Volume 56)
8. Anionic Surfactants: Chemical Analysis, *edited by John Cross* (out of print)
9. Stabilization of Colloidal Dispersions by Polymer Adsorption, *Tatsuo Sato and Richard Ruch* (out of print)
10. Anionic Surfactants: Biochemistry, Toxicology, Dermatology, *edited by Christian Gloxhuber* (see Volume 43)
11. Anionic Surfactants: Physical Chemistry of Surfactant Action, *edited by E. H. Lucassen-Reynders* (out of print)
12. Amphoteric Surfactants, *edited by B. R. Bluestein and Clifford L. Hilton* (see Volume 59)
13. Demulsification: Industrial Applications, *Kenneth J. Lissant* (out of print)
14. Surfactants in Textile Processing, *Arved Datyner*
15. Electrical Phenomena at Interfaces: Fundamentals, Measurements, and Applications, *edited by Ayao Kitahara and Akira Watanabe*
16. Surfactants in Cosmetics, *edited by Martin M. Rieger* (out of print)
17. Interfacial Phenomena: Equilibrium and Dynamic Effects, *Clarence A. Miller and P. Neogi*
18. Surfactant Biodegradation: Second Edition, Revised and Expanded, *R. D. Swisher*
19. Nonionic Surfactants: Chemical Analysis, *edited by John Cross*
20. Detergency: Theory and Technology, *edited by W. Gale Cutler and Erik Kissa*

21. Interfacial Phenomena in Apolar Media, *edited by Hans-Friedrich Eicke and Geoffrey D. Parfitt*
22. Surfactant Solutions: New Methods of Investigation, *edited by Raoul Zana*
23. Nonionic Surfactants: Physical Chemistry, *edited by Martin J. Schick*
24. Microemulsion Systems, *edited by Henri L. Rosano and Marc Clausse*
25. Biosurfactants and Biotechnology, *edited by Naim Kosaric, W. L. Caims, and Neil C. C. Gray*
26. Surfactants in Emerging Technologies, *edited by Milton J. Rosen*
27. Reagents in Mineral Technology, *edited by P. Somasundaran and Brij M. Moudgil*
28. Surfactants in Chemical/Process Engineering, *edited by Darsh T. Wasan, Martin E. Ginn, and Dinesh O. Shah*
29. Thin Liquid Films, *edited by I. B. Ivanov*
30. Microemulsions and Related Systems: Formulation, Solvency, and Physical Properties, *edited by Maurice Bourrel and Robert S. Schechter*
31. Crystallization and Polymorphism of Fats and Fatty Acids, *edited by Nissim Garti and Kiyotaka Sato*
32. Interfacial Phenomena in Coal Technology, *edited by Gregory D. Botsaris and Yuli M. Glazman*
33. Surfactant-Based Separation Processes, *edited by John F. Scamehom and Jeffrey H. Harwell*
34. Cationic Surfactants: Organic Chemistry, *edited by James M. Richmond*
35. Alkylene Oxides and Their Polymers, *F. E. Bailey, Jr., and Joseph V. Koleske*
36. Interfacial Phenomena in Petroleum Recovery, *edited by Norman R. Morrow*
37. Cationic Surfactants: Physical Chemistry, *edited by Donn N. Rubingh and Paul M. Holland*
38. Kinetics and Catalysis in Microheterogeneous Systems, *edited by M. Grätzel and K. Kalyanasundaram*
39. Interfacial Phenomena in Biological Systems, *edited by Max Bender*
40. Analysis of Surfactants, *Thomas M. Schmitt*
41. Light Scattering by Liquid Surfaces and Complementary Techniques, *edited by Dominique Langevin*
42. Polymeric Surfactants, *Irja Piirma*
43. Anionic Surfactants: Biochemistry, Toxicology, Dermatology. Second Edition, Revised and Expanded, *edited by Christian Gloxhuber and Klaus Künstler*
44. Organized Solutions: Surfactants in Science and Technology, *edited by Stig E. Friberg and Björn Lindman*
45. Defoaming: Theory and Industrial Applications, *edited by P. R. Garrett*
46. Mixed Surfactant Systems, *edited by Keizo Ogino and Masahiko Abe*
47. Coagulation and Flocculation: Theory and Applications, *edited by Bohuslav Dobiáš*
48. Biosurfactants: Production • Properties • Applications, *edited by Naim Kosaric*
49. Wettability, *edited by John C. Berg*
50. Fluorinated Surfactants: Synthesis • Properties • Applications, *Erik Kissa*
51. Surface and Colloid Chemistry in Advanced Ceramics Processing, *edited by Robert J. Pugh and Lennart Bergström*

52. Technological Applications of Dispersions, *edited by Robert B. McKay*
53. Cationic Surfactants: Analytical and Biological Evaluation, *edited by John Cross and Edward J. Singer*
54. Surfactants in Agrochemicals, *Tharwat F. Tadros*
55. Solubilization in Surfactant Aggregates, *edited by Sherril D. Christian and John F. Scamehorn*
56. Anionic Surfactants: Organic Chemistry, *edited by Helmut W. Stache*
57. Foams: Theory, Measurements, and Applications, *edited by Robert K. Prud'-homme and Saad A. Khan*
58. The Preparation of Dispersions in Liquids, *H. N. Stein*
59. Amphoteric Surfactants: Second Edition, *edited by Eric G. Lomax*
60. Nonionic Surfactants: Polyoxyalkylene Block Copolymers, *edited by Vaughn M. Nace*
61. Emulsions and Emulsion Stability, *edited by Johan Sjöblom*

ADDITIONAL VOLUMES IN PREPARATION

Applied Surface Thermodynamics, *edited by A. W. Neumann and Jan K. Spelt*

Vesicles, *edited by Morton Rosoff*

Liquid Detergents, *edited by Kuo-Yann Lai*

NONIONIC SURFACTANTS

Polyoxyalkylene Block Copolymers

edited by
Vaughn M. Nace
The Dow Chemical Company
Freeport, Texas

CRC Press is an imprint of the
Taylor & Francis Group, an **informa** business

CRC Press
Taylor & Francis Group
6000 Broken Sound Parkway NW, Suite 300
Boca Raton, FL 33487-2742

First issued in paperback 2019

ISBN-13: 978-0-8247-9700-3 (hbk)
ISBN-13: 978-0-367-40136-8 (pbk)

Visit the Taylor & Francis Web site at
http://www.taylorandfrancis.com

and the CRC Press Web site at
http://www.crcpress.com

Library of Congress Cataloging-in-Publication Data

Nonionic surfactants: polyoxyalkylene block copolymers / edited by
Vaughn M. Nace.
p. cm. – (Surfactant science series ; v. 60)
Includes index.
ISBN 0-8247-9700-0 (hardcover: alk. paper)
1. Surface active agents. 2. Block copolymers. I. Nace, Vaughn
Mark. II. Series.
TP994.N66 1996
668'.1–dc20 96-2619
CIP

Preface

Polyoxyalkylene block copolymer surfactants are a diverse subset of nonionic surfactants. This diversity results from the numerous structural possibilities available to the oxide chemist during synthesis. This volume focuses mainly on polyoxyalkylene copolymers made from the appropriate sequential block polymerization of ethylene oxide (EO), propylene oxide (PO), and 1,2-butylene oxide (BO). Some discussion of polyoxyalkylene homopolymer precursors of block copolymers is presented, when necessary, to provide clarification or explanation. Because this subject is treated fully in Volume 1 of this series, we have not focused heavily on surfactants made from hetero-polymerized or randomly polymerized oxides. Block copolymers derived from other monomers such as styrenics are not discussed because many tertiary references are already available.

The volume will be useful not only to researchers and application chemists working in the area of polyoxyalkylene surfactants, but also to those who develop closely related materials such as alkyl phenol ethoxylates, fatty alcohol ethoxylates, and other alkoxylated compounds.

Polyoxyalkylene block copolymers used in the urethanes industry are typically trifunctional and higher. Moreover, they are not used as surfactants per se, but as an integral chemical building block of the urethane system. Polymers having hydroxyl functionality greater than two will be discussed only in relation to their potential use as nonionic surfactants or in synthetic procedures.

Polyoxyalkylene copolymers are unique in that several degrees of freedom found in the manufacturing process may be utilized to give molecules that are tailor-made for specific uses. Among these are the variables of initiator type, initiator functionality, oxide species, oxide feed order, and block molecular weight. With these degrees of synthetic freedom in mind, Chapter 1 provides a thorough discussion of oxide polymerization fundamentals. Much of the information is offered as a framework for real-life synthetic situations, clearly emphasizing the aspect of block copolymer purity in terms of synthetic procedures. Chemical modification of the hydroxyl end group is also discussed.

Chapter 2 covers modern methods for the chemical analysis of polyoxyalkylene block copolymers. A thorough, well-referenced background in block copolymer analysis is presented. Copolymer structural analysis is then discussed in four sections covering general characterization, molecular weight, EO/PO ratio, and unsaturation. Analytical details for measuring copolymer composition as a whole are also given.

Chapter 3 presents an in-depth view of the physical chemical aspects of polyoxyalkylene block copolymers. The dilute aqueous association characteristics of block copolymer surfactants are discussed in terms of critical micelle concentration, critical micelle temperature, micellization thermodynamics, micellar structure, and association number. The effects of block copolymer nonhomogeneity and block architecture are also presented in detail. Block copolymer surface activity and solubilization phenomena are fully documented by using extensive tabulated data derived from numerous sources. The unique phase behavior and gelation properties of some polyoxyalkylene block copolymers have been the topics of numerous research manuscripts covering a 20 year period, and accelerating. The last half of Chapter 3 covers these aspects in great detail.

Chapter 4 deals with the properties of polyoxyalkylene block copolymers as related to physical handling and surfactant performance. The properties of crystallinity, melt viscosity, hydrophobe polarity, and thermal stability are discussed. Generalized wetting and foaming data are presented followed by a section giving detailed information on the relative performance of block copolymers based on polyoxypropylene and polyoxybutylene hydrophobes.

Chapter 5 is an up-to-date synopsis of applications for polyoxyalkylene block copolymer surfactants. Numerous examples are included covering the areas of medicine, coal and petroleum, plastics, emulsion polymerization, paper, photography, cleaner systems, personal care, and others.

Toxicological properties of polyoxyalkylene block copolymers are summarized in Chapter 6. Short- and long-term mammalian toxicity information along with metabolic data are provided. Mutagenicity and genotoxicity is discussed in conjunction with the subject of human safety followed by a section on structure–activity relationships.

Chapter 7, the final chapter, but by no means the least important chapter, treats environmental aspects of block copolymers. Biodegradation and aquatic toxicity are important topics for any modern discussion of surfactant materials. The chapter covers a variety of useful subjects including the environmental distribution of block copolymers, aquatic toxicity, and biodegradability issues including analytical methods, and degradation mechanisms.

I want to thank Dr. Martin J. Schick, who continues to be the driving force and visionary for the Surfactant Science Series. His help in obtaining the best chapter authors for this book was invaluable. I would also like to thank Dr. Irving R. Schmolka for his moral support and for showing his deep understanding of this subject. The teachings of Milton J. Rosen have been invaluable to me and others in the surfactant science field; my sincere gratitude goes to him. A multitude of thanks goes to the individual authors for their time and effort in putting together their respective chapters. Lastly, I would like to thank Richard E. Rozelle of The Dow Chemical Company for his support in the technical production of this work.

Vaughn M. Nace

Contents

Preface iii
Contributors ix

1. Synthesis and Chemical Modification of Polyoxyalkylene Block Copolymers 1
Robert H. Whitmarsh

2. Chemical Analysis of Polyoxyalkylene Block Copolymers 31
Henry T. Kalinoski

3. Physical Chemistry of Polyoxyalkylene Block Copolymer Surfactants 67
Benjamin Chu and Zukang Zhou

4. Properties of Polyoxyalkylene Block Copolymers 145
Vaughn M. Nace

5. Applications of Polyoxyalkylene Block Copolymer Surfactants 185
Michael W. Edens

6. Toxicology of Polyoxyalkylene Block Copolymers 211
Stephen C. Rodriguez and Edward J. Singer

7. Biological Activity of Polyoxyalkylene Block Copolymers in the Environment 243
Robert E. Bailey

Index 259

Contributors

Robert E. Bailey[*] Environmental Toxicology Laboratory, The Dow Chemical Company, Midland, Michigan

Benjamin Chu Department of Chemistry, State University of New York at Stony Brook, Stony Brook, New York

Michael W. Edens Industrial Polyglycol Research and Development, The Dow Chemical Company, Freeport, Texas

Henry T. Kalinoski Analytical Chemistry Section, Unilever Research U.S., Edgewater, New Jersey

Vaughn M. Nace Industrial Polyglycol Research and Development, The Dow Chemical Company, Freeport, Texas

Stephen C. Rodriguez[†] Department of Toxicology, Stonybrook Laboratories Inc., Princeton, New Jersey

---

Current affiliations:
[*] Bailey Associates, Midland, Michigan
[†] Central Research, Rhône-Poulenc Rorer, Collegeville, Pennsylvania

Edward J. Singer Technical Consultant, Environmental Science & Toxicology, Belle Mead, New Jersey

Robert H. Whitmarsh Industrial Polyglycol Research and Development, The Dow Chemical Company, Freeport, Texas

Zukang Zhou Department of Chemistry, State University of New York at Stony Brook, Stony Brook, New York

1

Synthesis and Chemical Modification of Polyoxyalkylene Block Copolymers

ROBERT H. WHITMARSH Industrial Polyglycol Research and Development, The Dow Chemical Company, Freeport, Texas

I. Polyether Overview 2
- A. General chemistry 2
- B. Molecular weight (block size) considerations 4
- C. Safety considerations 6

II. Preparation and Characterization of Block Copolymers 7
- A. Polymer initiators 7
- B. Base catalyst 7
- C. Base neutralization 8
- D. Polyether structural considerations 9
- E. Planning the synthesis 11
- F. Polymer characterization 12

III. Synthetic Challenges 13
- A. Deviations from Flory's first and second assumptions 14
- B. Deviations from Flory's third assumption 16
- C. Other common synthetic problems 21

IV. Modification of Polyoxyalkylene Block Copolymers 22
- A. Modification of the hydroxyl group 22
- B. Modification of the polyether backbone 25

References 25

I. POLYETHER OVERVIEW

A. General Chemistry

Polyethers are those polymers containing ether (C-O-C) linkages in the chain backbone. Such polymers are quite numerous. They are derived from a wide variety of monomers, by many different synthetic routes. Polyethers find utility in diverse applications either directly or as chemical intermediates [1]. Of necessity this chapter deals with a limited subset of the total polyether realm. Only those polyethers derived from oxirane (also referred to as ethylene oxide and EO), methyloxirane (propylene oxide and PO), and ethyloxirane (1,2–butylene oxide and BO) will be discussed. Structures of these monomers are shown in Fig. 1. As a class these monomers are called oxiranes, alkylene oxides, AOs, or simply oxides.

While much of what will be said applies to all polyethers, the emphasis of this chapter follows the surfactant theme of the book and deals mainly with the synthesis and, to a lesser degree, the modification of block copolymers. Although many different catalyst systems are used for oxide polymerization [2], only the base-catalyzed mechanism is discussed here. The decision to focus only on preparations using base catalysis is appropriate, as strong base is the catalyst system used for most polyoxyalkylene block copolymers offered as surfactants by industry today [3,4]. Other catalysts used today cover a range of compositions including Lewis acids [5–7], metal coordination catalysts [8–10], and metal porphyrin [11–13]. Some of these catalyst systems readily provide stereoregular polyalkylene glycols of PO and BO [14,15]. Base catalyst is able to do so only when pure *l* or *d* oxide isomers are used [16]. While this is an interesting area, no more will be said about stereoregular polyalkylene glycols in this chapter.

Various drawbacks are found in every catalyst system used to date. The reader is encouraged to consider alternate synthetic methods which overcome the problems associated not only with base catalyst discussed below but also

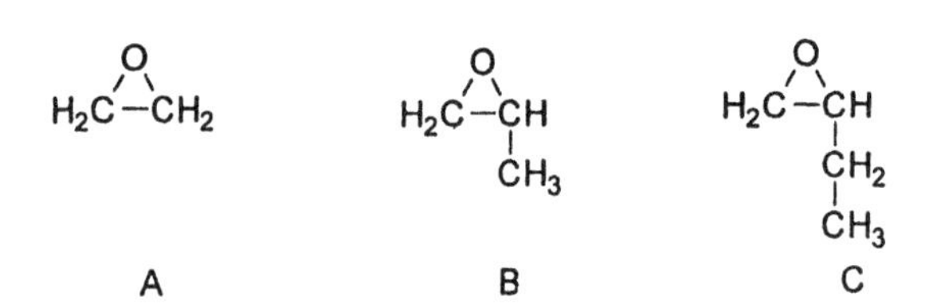

FIG. 1 Oxiranes of interest. A) Ethylene oxide. B) Propylene oxide. C) 1,2–Butylene oxide.

to become familiar with the other systems and the problems associated with each of them.

The base-catalyzed polymerization of alkylene oxides has been known for over a century [17,18]. Industrial applications emerged in the late 1940s [19,20]. Although this chapter is not intended to be all inclusive, it is the author's wish that it be of benefit to surfactant scientists. For those workers who are now discovering the many uses of block copolymers as surfactants the chapter will serve as an introduction to the fundamental chemistry of alkylene oxide polymers. Appreciation of the many examples of how synthesis can go awry will help all workers. Remembering these anomalies will help to explain erratic results from the measurement of surfactant behavior.

Oxiranes are heterocyclic compounds consisting of two carbon and one oxygen atoms in the ring. All such rings are readily opened by a number of reagents. However, many highly substituted oxiranes do not undergo polymerization or do so only with difficulty. At the other extreme EO, PO, and BO polymerize readily with either acid or base catalysis. Commercial polyethers of low molecular weight (less than 20,000 daltons), narrow molecular weight dispersity, and based on these three oxides are prepared using strong base to facilitate the reaction. Base-catalyzed polymerization of alkylene oxides involves nucleophilic attack of an initiator molecule on one carbon of the ring resulting in hetero-bond cleavage. This bond cleavage generates an alkoxide anion. Itself an excellent nucleophile, the alkoxide anion reacts with another oxide molecule, thus propagating polyether chain growth. This is a second-order nucleophilic replacement (S_N2) reaction [21]. The substitution reaction is useful for polymerization as the leaving group is not completely released from the molecule. In this way the oxiranes are bifunctional. Figure 2 depicts this propagation reaction; the formation of secondary hydroxyl groups (Reaction A) is greatly favored over primary hydroxyl groups (Reaction B) when propylene oxide and butylene oxide are polymerized under basic conditions [22]. This has been well known for many years. The implications of this selectivity will be discussed further below.

When low molecular weights are involved, this system appears to be a living polymerization [23]. This pseudoliving process allows ready preparation of block copolymers. This is accomplished by reacting different monomers sequentially. Discrete blocks are formed when the concentration of the first oxide is reduced to essentially zero before the introduction of the second oxide. A transition zone containing both monomer units results when the second oxide is introduced before the first has completely reacted. Polyethers with random distributions of monomer units are made by adding different oxides together [24]. Deviation from a true living polymerization is discussed later in the chapter.

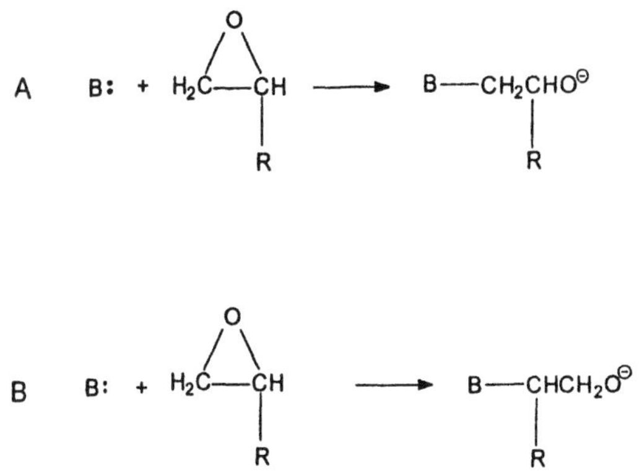

$$\text{A} \quad \text{B:} + \underset{\text{R}}{\text{H}_2\text{C}\overset{\text{O}}{—}\text{CH}} \longrightarrow \text{B—CH}_2\underset{\text{R}}{\text{CH}}\text{O}^{\ominus}$$

$$\text{B} \quad \text{B:} + \underset{\text{R}}{\text{H}_2\text{C}\overset{\text{O}}{—}\text{CH}} \longrightarrow \text{B—}\underset{\text{R}}{\text{CH}}\text{CH}_2\text{O}^{\ominus}$$

FIG. 2 Polymerization by SN2 propagation step. R = H, Me, or Et. Polymerizing ethylene oxide, R = H, reactions A and B are identical. With PO and BO, the reaction forming a secondary hydroxyl group (Reaction A) is greatly favored over formation of a primary hydroxy group (Reaction B). Secondary to primary ratio exceeds ten to one. *Source*: from [22].

Polyoxyalkylene block copolymers are efficient surfactants. In these materials EO block(s) provide hydrophilicity and PO (or BO) block(s) the hydrophobicity necessary for surfactancy [25,26].

B. Molecular Weight (Block Size) Considerations

Polymerization of oxiranes under basic conditions does not proceed indefinitely. Molecular weights of polyethers prepared with a strong base catalyst might approach 20,000 daltons if a molecule with multiple nucleophilic sites is used as the initiating species. With a single site, however, the molecular weight will seldom exceed 5,000 daltons [27,28]. This molecular weight limit is imposed chemically by extraneous introduction of molecules which initiate growth of new polymer chains [29]. From a practical standpoint, molecular weights for industrial processes are limited by increasingly smaller polymerization rates due to a decrease in catalyst concentration as the polymerization proceeds.

Molecular weight distributions in polyethers are generally very narrow. Polydispersities of 1.05 to 1.15 are typical for polyethylene oxides [21] and slightly wider for polypropylene and polybutylene oxides [30]. Narrow dispersity results from three important characteristics of the reacting system:

- all molecules have an equal opportunity to react with monomer,
- reaction rates of addition are equal regardless of molecular weight,

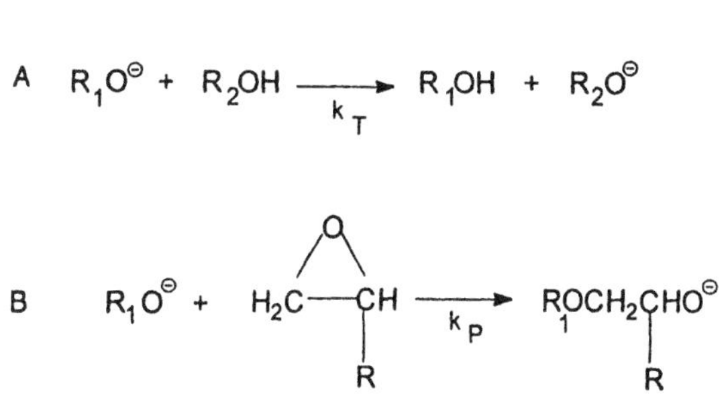

$$\text{A} \quad R_1O^{\ominus} + R_2OH \xrightarrow[k_T]{} R_1OH + R_2O^{\ominus}$$

$$\text{B} \quad R_1O^{\ominus} + H_2C\overset{O}{\frown}CH(R) \xrightarrow[k_P]{} R_1OCH_2CH(R)O^{\ominus}$$

FIG. 3 Reactions of alkoxide anions. A) Acid-base chain transfer reaction. B) Chain propagation reaction. When EO, PO, and BO reacts with alkoxides $k_T >> k_P$.

- all initiators are present from the beginning of the polymerization.

Alkoxide ions are the nucleophiles which extend the polyether chain. On a molar basis the alkoxide ions represent a small fraction of the molecules present. This fraction could be less than 0.001 or greater than 0.01. A fraction of 0.1 would be very high. In a low molecular weight alcohol a mole fraction of alkoxide ions equal to 0.1 would increase the polymerization rate such that control would be extremely difficult. In a high molecular weight alcohol, such as a hydroxyl-terminated polyether, great difficultly would be encountered in trying to dissolve the necessary amount of strong base to generate a fraction of alkoxide as high as 0.1.

In a system where the true reactive species represents a small fraction of all molecules, why is it that the final distribution is not one containing a few large molecules and a great number of low molecular weight oligomers, even initiator molecules? The answer is the alkoxide ion undergoes two different reactions (Fig. 3). The first reaction can be described as an acid-base chain transfer reaction of alkoxide ion with an alcohol. The second reaction consists of a propagation step, the addition of an oxirane monomer to the alkoxide ion, and increasing chain length by one unit. Investigation of the rate constants k_p and k_T shows why the molecular weight distribution (mwd) is small. Transfer rates are much greater than propagation rates so that at any instant all of the molecules present have an equal chance of being in the alkoxide state [31].

A second feature of alkylene oxide polymerization reactions is that the reactivity of all alkoxide ions is essentially equal regardless of molecular weight (Fig. 4) [21,31,32].

The final feature of the polymerizing system leading to narrow molecular weight distributions is that all initiator molecules are present at the first introduction of alkylene oxide. Thus, each polymer chain will encounter, on average, the same number of monomers.

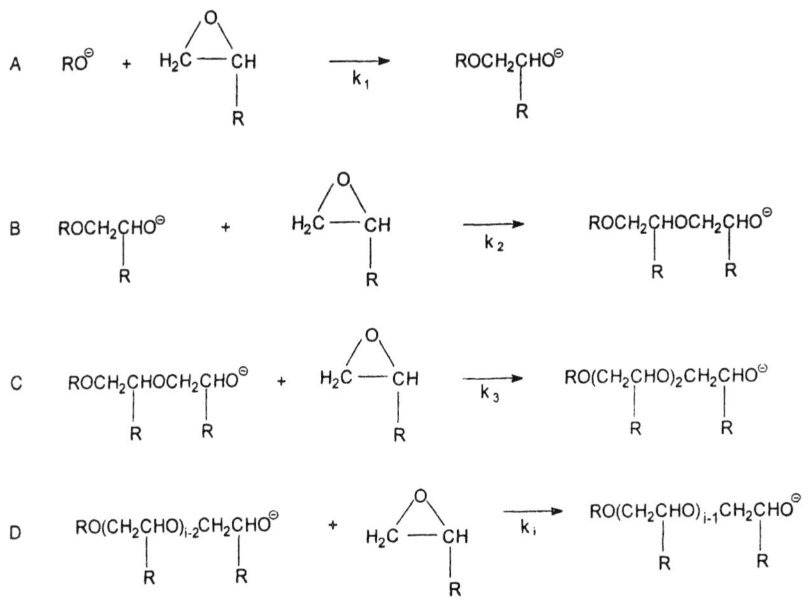

A $$RO^{\ominus} + H_2C\overset{O}{—}CH(R) \xrightarrow{k_1} ROCH_2CH(R)O^{\ominus}$$

B $$ROCH_2CH(R)O^{\ominus} + H_2C\overset{O}{—}CH(R) \xrightarrow{k_2} ROCH_2CH(R)OCH_2CH(R)O^{\ominus}$$

C $$ROCH_2CH(R)OCH_2CH(R)O^{\ominus} + H_2C\overset{O}{—}CH(R) \xrightarrow{k_3} RO(CH_2CH(R)O)_2CH_2CH(R)O^{\ominus}$$

D $$RO(CH_2CH(R)O)_{i-2}CH_2CH(R)O^{\ominus} + H_2C\overset{O}{—}CH(R) \xrightarrow{k_i} RO(CH_2CH(R)O)_{i-1}CH_2CH(R)O^{\ominus}$$

FIG. 4 Propagation rate constants. Consecutive reactions of initiating alkoxide with monomer units. A) First monomer addition. B) Second monomer addition. C) Third monomer addition. D) i^{th} monomer addition. Ideally $k_1 = k_2 = k_3 = \ldots = k_i$. In actuality, $k \approx k_2 \approx k_3 \approx \ldots \approx k_i$.

Taken together these three conditions (i.e., equal opportunity for all molecules to react, constant reaction rates regardless of molecular weight, and all polymer chains started at the same time) lead to a Poisson-type distribution of molecular weights [33]. This distribution is exemplified by polydispersities close to one. While measured polydispersities are small, they are never equal to the to predicted Poisson distribution [34–36]. This is due to none of the three statements ever being completely satisfied [37–39]. A large part of the remaining chapter describes why polymerizations differ from these three conditions in actual experience. While never completely satisfied, these three statements nonetheless can be used effectively as simplifying assumptions to predict closely the actual values of each molecule's chances of adding monomer, and thereby the average molecular weight and molecular weight distribution.

C. Safety Considerations

Appropriate equipment for working with alkylene oxides and an understanding of the precautions which must be rigorously followed are absolutely mandatory to safely perform these syntheses. This chapter is not intended to address all concerns in carrying out this chemistry. However, it must always be remembered that oxide monomers may exhibit both immediate and long-term toxicity. Even brief, low level exposure may result in detrimental health effects. Current Material Safety Data Sheets and other safe handling data must be consulted for up-to-date information on the use of alkylene oxides. Alkylene oxide monomers are flammable and reactive chemicals. Generally, AO work is done only in steel vessels capable of containing pressures exceeding 100 psig, with sufficient heat removal capability to handle the considerable heat of reaction during polymerization, and equipped with adequate means of safely relieving pressure should a run away reaction occur.

II. PREPARATION AND CHARACTERIZATION OF BLOCK COPOLYMERS

A. Polymer Initiators

As just discussed, the synthesis of polyethers requires an initial attack of a nucleophile on an oxirane; this generates an alkoxide that is the nucleophile for the next monomer addition. Phenols, carboxylic acids, amines, alcohols, and even water are sufficiently nucleophilic to undergo the first reaction without the use of catalyst (40–42). Using the noncatalyzed nucleophiles, chain propagation is very limited because reaction rates are low. Seldom do these chains add more than two or three monomer units. The alkoxide ions react rapidly with oxiranes but in most cases the ions formed are almost instantaneously neutralized internally. Thus there is little, if any, chain extension.

Only the amines are capable of maintaining alkoxide ions for appreciable lengths of time. This occurs because quaternary ammonium ions are present to stabilize the alkoxide anion charge [43–45]. Initially, zwitterions are formed by the addition of alkylene oxide to an amine nitrogen (Fig. 5A). As long as there is an amine hydrogen present, the zwitterion rapidly converts to the more thermodynamically stable amino alcohol (Fig. 5B). This transition is analogous to the internal neutralization of other starter molecules. A relatively stable zwitterion is formed after all amine hydrogens are replaced (Fig. 5C) [46,47]. Longer, yet still molecular weight limited, polyether chains can be prepared in this way. Molecular weight limitations are due to Hoffman elimination from the quaternary amine [48].

B. Base Catalyst

To achieve higher molecular weights, the presence of alkoxide ion must be maintained. This is accomplished by addition of strong base to the system [49].

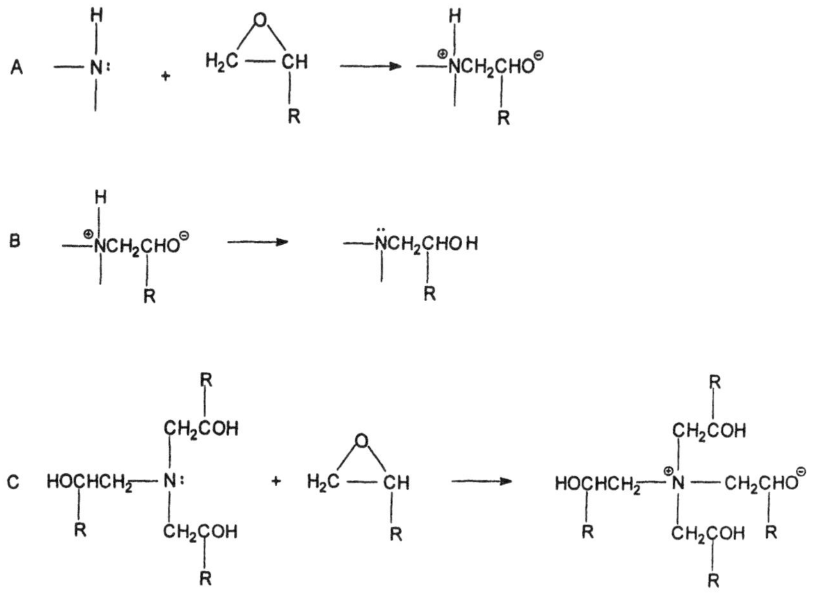

FIG. 5 Formation and fate of zwitterions. A) Zwitterion formation by reaction of amine with alkylene oxide. B) Neutralization to amino alcohol. C) Formation of quaternary zwitterion available for catalyzing polymerization or Hoffman elimination.

Most commonly the base is sodium or potassium hydroxide. Methoxides and ethoxides of these same metals are encountered, but more infrequently. The use of the metals themselves is limited, and mainly exclusive to the laboratory.

With one of these strong bases present, all of the nucleophiles previously mentioned can be used as initiators in preparing high molecular weight polyethers. While amines and carboxylic acids can be used, they seldom are unless there is an important need for their contributions to the final molecules. Amines always are plagued by the Hoffman elimination reaction which severely limits molecular weight and broadens the distribution [50,51]. Carboxylic acids are susceptible to transesterification, leading to undesirable mixtures of diesters and diols [52,53]. Polyethers initiated with phenols, alcohols, or water are most common because the growing chains are not subject to undesirable side reactions during synthesis.

C. Base Neutralization

While the introduction of one of the sodium or potassium bases is necessary for preparation of polyethers, for most practical applications the base must be neutralized or otherwise removed from the polymeric product [54–56]. Neutralization with an acid is accompanied by the formation of alkali metal salt; for surfactant use, this salt may need to be removed. A decision to leave or remove the salt to an acceptable level is based both on economic and performance considerations. Little more will be said concerning neutralization and salt removal.

The presence of base, acid, or salt can affect surfactant properties such as cloud point, surface tension, viscosity of solutions, and critical micelle concentrations (cmc) [57–61]. The researcher is advised to characterize all samples carefully for over- or under-neutralization (i.e., acidic or basic polymer) and the presence of salt. Manufacturers use different processes at the point of neutralization and normal process variation may allow for rather large variations in salt levels. Uncontrolled and, perhaps worse, unknown variation in this area can invalidate otherwise carefully constructed and executed experiments.

D. Polyether Structural Considerations

Polyoxyalkylene block copolymers possessing the appropriate hydrophile-lipophile balance are effective surfactants in a wide range of applications. The synthetic chemist can manipulate block size and arrangement to achieve desired surfactant performance. Also at their disposal is the choice of initiator. Two considerations are applied in the selection of an initiator.

The first consideration is the influence of initiator type on block copolymer surfactancy. The initiator can have a variety of functional groups, many of which are polar. Even with only the simple hydroxyl function, the initiator can influence surfactant properties. Consider surfactants with identical block size and arrangement. It is not surprising that a surfactant initiated from methanol differs from one initiated with a mixture of branched C_{16} and C_{18} alcohols. Likewise, there is a noted difference between a surfactant initiated with a linear long-chain alcohol compared to a highly branched alcohol with identical carbon number [62]. In addition to the size and hydrophobicity of the initiator, the synthetic chemist can choose initiators with multiple hydroxyl functionality and thereby have the ability to increase the number of polyether chains per molecule and increase the polymer molecular weight.

Inevitable side reactions of PO and BO limit the maximum chain length per hydroxyl group and therefore limit the maximum hydroxyl equivalent weight (HEW) [63], which will be discussed in detail later. Higher molecular weight polymers can be prepared using initiators of greater functionality [24,30]. To the polyether chemist, functionality refers to the number of hydroxyl groups per molecule. The functionality of a polyether stems from the initiator molecule which began the polymerization. Figure 6 illustrates polyethers initiated by methoxide, hydroxide, and 1,2,3–propanetriol, monosodium salt. The polyether

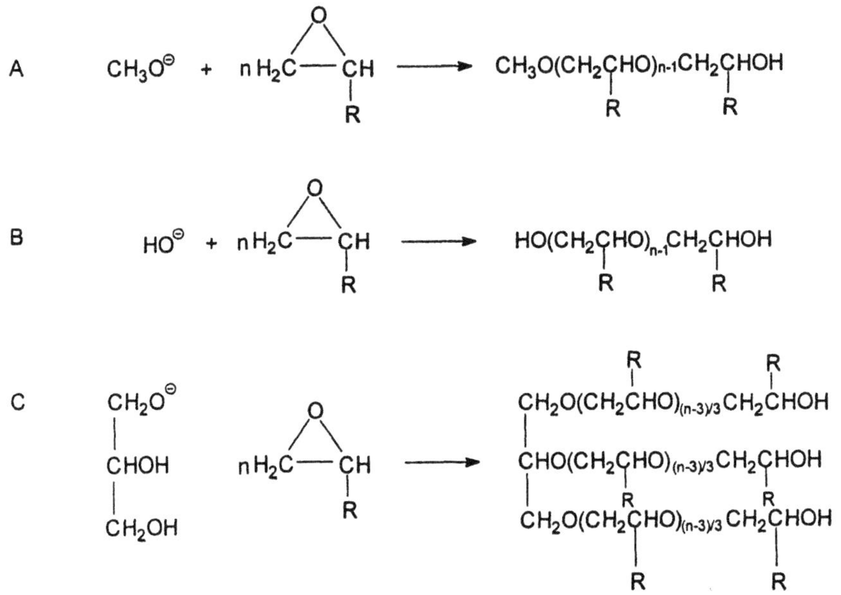

A $CH_3O^{\ominus} + n\,H_2C\overset{O}{—}CH(R) \longrightarrow CH_3O(CH_2CH(R)O)_{n-1}CH_2CH(R)OH$

B $HO^{\ominus} + n\,H_2C\overset{O}{—}CH(R) \longrightarrow HO(CH_2CH(R)O)_{n-1}CH_2CH(R)OH$

C $CH_2O^{\ominus}—CHOH—CH_2OH + n\,H_2C\overset{O}{—}CH(R) \longrightarrow$

$CH_2O(CH_2CH(R)O)_{(n-3)/3}CH_2CH(R)OH$

$CHO(CH_2CH(R)O)_{(n-3)/3}CH_2CH(R)OH$

$CH_2O(CH_2CH(R)O)_{(n-3)/3}CH_2CH(R)OH$

FIG. 6 Polyether functionality. A) Monol. B) Diol. C) Triol. In the 1,2,3–propanetriol molecule only one hydroxyl is shown as an alkoxide anion, yet all three hydroxyls react. This is because the chain transfer reaction is much more rapid than chain extension and, therefore, all three hydroxyl groups are equally reactive.

products are monol, diol, and triol respectively. Functionality higher than three is available, if desired.

Initiator functionality also determines the number of blocks present following addition of two or more different alkylene oxides. Generally, the number of blocks is taken to be equal to the product of functionality and number of separate oxide additions. For small initiators, a better picture of the polyoxyalkylene block copolymer is provided by considering the polymer after one oxide addition as a central core unit, regardless of functionality. In this way the number of blocks after adding two AOs is equal to functionality plus one. This is shown graphically in Fig. 7. With continued addition of oxiranes the number of blocks increases according to the formula:

$$B = 1 + (n - 1) \times F$$

where B is the number of blocks, n is the number of different oxide additions, and F is the initiator functionality. Keep in mind that this calculation assumes that the initiator is small enough that the first AO addition creates only one core block, regardless of the functionality.

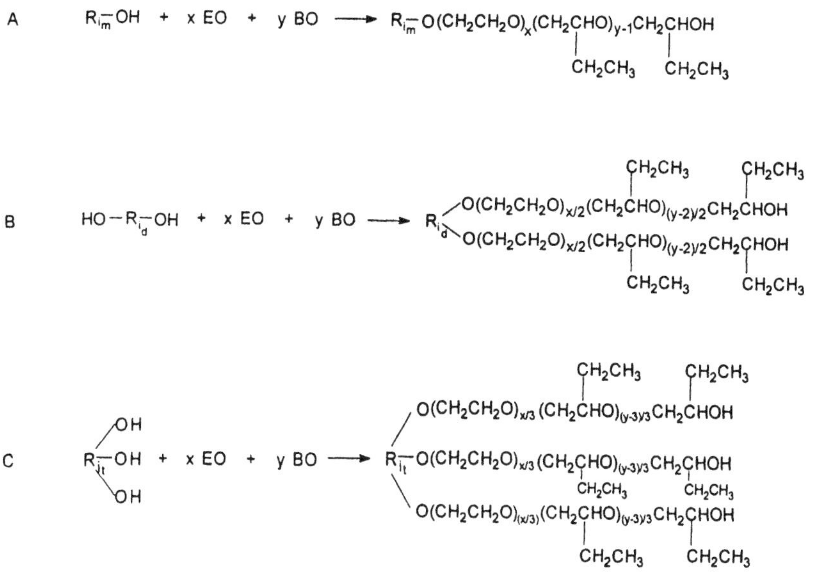

A $R_{i_m}\text{–OH} + x\ \text{EO} + y\ \text{BO} \longrightarrow R_{i_m}\text{–O}(CH_2CH_2O)_x(CH_2CH(CH_2CH_3)O)_{y-1}CH_2CH(CH_2CH_3)OH$

B $HO\text{–}R_{i_d}\text{–OH} + x\ \text{EO} + y\ \text{BO} \longrightarrow R_{i_d}[\text{O}(CH_2CH_2O)_{x/2}(CH_2CH(CH_2CH_3)O)_{(y-2)/2}CH_2CH(CH_2CH_3)OH]_2$

C $R_{i_t}(\text{OH})_3 + x\ \text{EO} + y\ \text{BO} \longrightarrow R_{i_t}[\text{O}(CH_2CH_2O)_{x/3}(CH_2CH(CH_2CH_3)O)_{(y-3)/3}CH_2CH(CH_2CH_3)OH]_3$

FIG. 7 Number of blocks depends on functionality. A) Monol, Ri_m, after two feeds contains two blocks. B) Diol, Ri_d, after two feeds contains three blocks. C) Triol, Ri_t, after two feeds contains four blocks. In general $B = 1 + (n - 1) \times F$. Variables are defined in the text. This is true provided that the first oxide feed forms a central core of one block, regardless of functionality.

E. Planning the Synthesis

Planning for the synthesis is straight forward. For a given problem, the chemist makes a prediction of structural requirements such as block size, block arrangement, initiator size and initiator functionality. The basis of the prediction might be literature reports, discussions with colleagues, personal experience or a combination of all three. Choice of base can also affect the copolymer product [64–66]. Depending on one's preference, the amounts of each component needed can be found by calculating backward using the desired amount of product to be made, or, forward from the amount of initiator on hand, or, from the amount of initiator necessary to achieve minimum charge to the available reactor system.

Consider the diblock monol initiated with *p*-octylphenol containing first a block of twenty moles of ethylene oxide followed by a block of eight moles of butylene oxide (Fig. 8). The component molecular weights are octylphenol, 206; EO, 44; and BO, 72. Summing the molecular weights of all components provides the number average molecular weight of the target surfactant. For this example, the sum is $206 + 20(44) + 8(72) = 1662$.

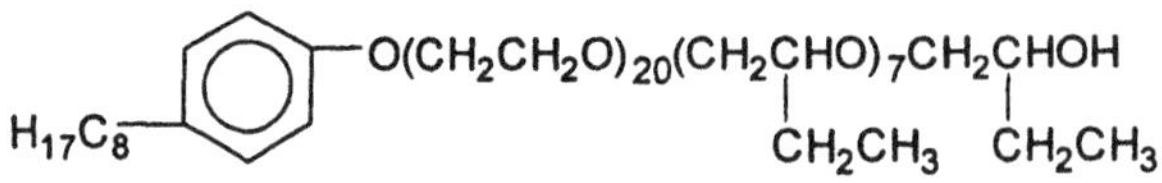

$H_{17}C_8-C_6H_4-O(CH_2CH_2O)_{20}(CH_2CH(CH_2CH_3)O)_7CH_2CH(CH_2CH_3)OH$

FIG. 8 An example of a block copolyether.

The required amounts of EO and BO are easily determined from the moles of octylphenol initiator present and molar block size desired. The following example uses 800 g of initiator:

$$800 \text{ g} \times 1 \text{ eq. OH}/206 \text{ g} = 3.88 \text{ eq. OH}$$

$$3.88 \text{ eq. OH} \times 20 \text{ moles EO/eq. OH} \times 44 \text{ g/mole} = 3414 \text{ g EO}$$

$$3.88 \text{ eq. OH} \times 8 \text{ moles BO/eq. OH} \times 72 \text{ g/mole} = 2235 \text{ g BO}$$

Similarly if 5,000 g of this surfactant is required for evaluation in a particular application, the polymer's number average molecular weight is used to find the number of moles octylphenol needed, and from that value, the masses of initiator, EO, and BO are found. Each mole of polymer is the result of one mole of initiator [24]. Therefore, knowing the moles of polymer required provides all the information necessary to find the moles of each component.

$$5000 \text{ g} \times 1 \text{ eq. OH}/1662 \text{ g} = 3.01 \text{ eq. OH}$$

$$3.01 \text{ eq. OH} \times 206 \text{ g/eq. OH} = 620 \text{ g octylphenol}$$

$$3.01 \text{ eq. OH} \times 20 \text{ moles EO/eq. OH} \times 44 \text{ g/mole} = 2649 \text{ g EO}$$

$$3.01 \text{ eq. OH} \times 8 \text{ moles BO/eq. OH} \times 72 \text{ g/mole} = 1734 \text{ g BO}$$

Extraneous initiation, either from contamination of the raw materials or side reactions, will affect the final molecular weight. The choice of sodium or potassium base influences the amount of extraneous initiation. These problems are discussed in more detail below.

F. Polymer Characterization

The final responsibility of the synthetic chemist is to confirm that the finished polymer is in fact the one originally conceived. The applications chemist should be concerned with the polymer characterization process especially when the surfactant has been obtained from an outside source. Characterization is not always a simple task due to the complexity of the polyoxyalkylene block copolymer mixture. In addition to molecular weight distribution common to all synthetic polymers, there are often molecules from side reactions and impurities [24,67]. Molecules resulting from extraneous initiation can differ greatly in size, functionality, and number of blocks from the desired surfactant molecule. Due to lack of understanding of side reactions, these components of the polymer mixture are often unnoticed. Differences between individual lots of surfactant may never be completely identified. However, one should not be taken by surprise when there are performance differences between lots which were prepared in "exactly the same way." Methods used to study subtle differences between similar copolymers include high performance liquid chromatography [68–70], size exclusion chromatography [71], thin layer chromatography [72], supercritical fluid chromatography [36,73–75], infrared spectroscopy [76], and proton nuclear magnetic resonance [77].

Complete characterization requires extensive use of instrumental and chromatographic techniques. The assistance of an analytical chemist with polymer experience is also valuable. The purpose here is not to delve deeply into methods for determining composition; rather it is to be a reminder that a polyoxyalkylene block copolymer surfactant is a very complex mixture [48,78–80]. Small changes during synthesis can have profound effects on the final polymer composition [81].

For polyoxyalkylene block copolymers prepared in-house, it is mandatory that deviation between expected and measured hydroxyl equivalent weight be explained. This deviation is a clear signal that the intended ratio of reactants was not achieved. The oxide-to-initiator ratio can be incorrect due to mistakes in adding the oxide, misweighing the initiator, or significant extraneous initiation. In some circumstances, the polymerization details are unavailable. When such is the case, the polymers themselves are certainly worthy of study, but results should be regarded with added skepticism.

III. SYNTHETIC CHALLENGES

Many problems may be encountered during synthesis which can affect important copolymer properties such as molecular weight (mw), mwd, functionality, block size, and number of blocks. Changes in any one of these factors results in a polyoxyalkylene block copolymer differing from the target polymer.

In this section the impact of each factor on block copolymer composition is discussed along with implications for surfactant use. While measured polydispersities are small, they are not equal to the predicted Poisson distribution because none of the simplifying assumptions used to calculate the molecular weight distribution are completely satisfied during polymerization. In actual polymerizations, all molecules might not have an equal likelihood of being in the alkoxide ion state due to poor mixing or differences in the acidity of the alcohols present. While normally of minor importance during polymerization, acidity is a factor which often causes the rate of addition of the first monomer to differ significantly from the second. This is especially true when the initiator is a phenol or when the hydroxyl of the initiator differs from that after the first oxide addition (i.e., a primary alcohol initiator polymerized with PO or BO, or a secondary alcohol initiator polymerized with EO) [43,82,83]. Propagation rates then change very slowly with increasing molecular weight [81,84,85]. Even when all alcohols have equal probability of being in the alkoxide form, primary and secondary alkoxides have different reaction rates due to steric differences [86,87]. This factor also effects on the block transition zone of a copolymer [26,27]. Finally, it is not possible to have a constant amount of initiator present throughout the polymerization. In all real systems, extraneous initiators are added continuously during the polymerization.

For many block copolymers, deviations from the ideal can be observed only by very careful analysis [88,89; see also 69–77]. The researcher should not forget how and why the product can differ from that intended. When the measured hydroxyl equivalent weight (HEW) is near the target, more assurance exists that the synthesis has been successful.

Deviations from the ideal Poisson distribution will be examined next by seeing how the actual chemistry differs from the assumptions used in deriving the molecular weight distribution. Such complicating factors can cause lot-to-lot and supplied-to-supplier variation.

A. Deviations from Flory's First and Second Assumptions

The first two assumptions, as stated by Flory, are that every molecule has equal opportunity to react and that reaction rates of every species present for reaction are equal [33–35]. Alcohols have unequal probability of existing as alkoxide ion because of differences in acidity of various hydroxyl groups. Phenols are much more acidic than alcohols. When phenols are used as initiators nearly every molecule will have added one oxide unit prior to any having added two units [31,90,91]. When 12–15 moles of oxide per mole of initiator have been added, the difference in molecular weight distributions between phenol and alcohol initiated polymers are minor [21]. Acidity differences of primary and secondary hydroxyl groups are slight and any difference in molecular weight distribution caused by this reason is relatively small.

The ideal of equal propagation rates is seldom, if ever, achieved in base-catalyzed polymerizations. Primary alkoxides react more readily with alkylene oxide than do secondary alkoxides. As shown in Fig. 9, the addition of PO or BO to primary alcohol initiator (Reaction A) is faster than an addition to a secondary alkoxide oligomer (Reaction B). Similarly, adding EO to a secondary alcohol initiator (Reaction C) is slower than addition to a primary alkoxide oligomer (Reaction D). As previously stated, hydroxyl acidity plays a minor

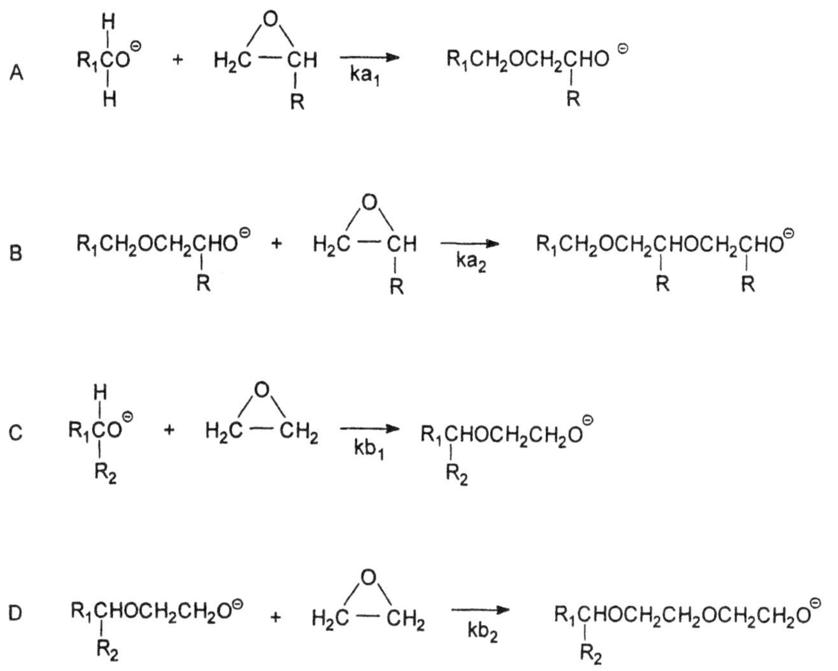

A $$R_1CH_2O^{\ominus} + H_2C\overset{O}{—}CH(R) \xrightarrow{ka_1} R_1CH_2OCH_2CH(R)O^{\ominus}$$

B $$R_1CH_2OCH_2CH(R)O^{\ominus} + H_2C\overset{O}{—}CH(R) \xrightarrow{ka_2} R_1CH_2OCH_2CH(R)OCH_2CH(R)O^{\ominus}$$

C $$R_1CH(R_2)O^{\ominus} + H_2C\overset{O}{—}CH_2 \xrightarrow{kb_1} R_1CH(R_2)OCH_2CH_2O^{\ominus}$$

D $$R_1CH(R_2)OCH_2CH_2O^{\ominus} + H_2C\overset{O}{—}CH_2 \xrightarrow{kb_2} R_1CH(R_2)OCH_2CH_2OCH_2CH_2O^{\ominus}$$

FIG. 9 Steric impact on hydroxyl activity. A) Substituted oxirane, PO or BO, addition to primary alkoxide ion generates a secondary alkoxide ion. Reaction rate is ka_1. B) Addition of substituted oxirane to secondary alkoxide ion formed by reaction A. Reaction rate is ka_2. For reactions A and B. $ka_1 > ka_2$. C) EO addition to secondary alkoxide ion generates a primary alkoxide ion. Reaction rate is kb_1. D) Addition of EO to primary alkoxide ion formed by reaction C. Reaction rate is kb_2. For reactions C and D, $kb_1 < kb_2$.

role in the rate differences. The overwhelming factor in rate differences with primary to secondary alkoxide ion is steric hinderance at the reactive site [92]. Steric hinderance reaches its extreme with 1,1,2,2–tetramethylethylene oxide where the polymerization step does not happen at all (Fig. 10) [63].

The transition from primary to secondary hydroxyl occurs in polyoxyalkylene block copolymers when changing from hydrophilic oxide (EO) to hydrophobic (PO or BO) oxide. A change from hydrophobic to hydrophilic oxide results in a change from secondary to primary hydroxyl end groups. The differing rates of addition in going from one to another alkylene oxide can cause significant property changes in the block copolymer product. Such changes are most dramatic when a short EO block is added to a block derived from PO or

$$(H_3C)_2C\overset{O}{—}C(CH_3)_2 + RO^{\ominus} \longrightarrow H_2C{=}C(CH_3)-C(CH_3)(OH)-CH_3 + ROH$$

FIG. 10 1,1,2,2–Tetramethylethylene oxide. No polymerization when reacted with strong base.

BO [26]. More or less efficient reaction of the EO with all of the molecules present has bearing on the distribution of individual EO block sizes but not average EO block size. Polyoxyalkylene block copolymers having the same overall composition can be either liquid (efficient capping of EO onto the hydrophobe) or solid (inefficient capping) [26]. With efficient EO capping, EO blocks are more closely clustered about an average size. Therefore, there are no extremely long blocks and melting points of this chain segment are below room temperature. Inefficient capping results in a wide range of chain lengths, some being much longer than the average block. Melting points of the long chains can be above room temperature [93]; consequently the copolymer is a paste or solid. Less efficient capping can affect surfactancy of the polymers as well. When some hydrophobic molecules receive large amounts of EO, others must receive very little, possibly none at all. The molecules with little or no EO block may not behave as a surfactant. This effect can confuse comparisons of cmc, surface tension, etc., of apparently identical polymers.

Even with no change in hydroxyl type, propagation rates tend to decrease very slowly with increasing molecular weight reaching a limiting value [81,84,94]. Regardless of the cause, changes in propagation rate generally lead to higher polydispersity. How the polydispersity change will affect surfactancy of block copolymers cannot be predicted. It may be of interest to the researcher to test polyethers which differ in polydispersity and EO capping efficiencies to determine effects in a specific application. It is worth reiterating that polyoxyalkylene block copolymers having the same average molecular weight and overall composition of blocks can behave differently in a particular system.

B. Deviations from Flory's Third Assumption

Extraneous initiation affects the composition of the product copolymer in several ways. The increase in molecular weight distribution as initiators are introduced during polymerization is an obvious change. More subtle are changes in average functionality. Also with extraneous initiation, not all molecules have the same number or size of blocks. Extraneous initiation results from two sources: impurities or side reactions [63,86,95].

$$ROH + Na^{\oplus}OH^{\ominus} \longrightarrow RO^{\ominus}Na^{\oplus} + H_2O$$

FIG. 11 Generation of extraneous diol initiator (water) with strong base and alcohol.

While other impurities can be present, the most common is water. When alkali metal hydroxides are used, water is generated along with the alkoxide [28] (Fig. 11). Unless the intended initiator is itself a diol, a mixture of differing functionality now exists. Another consideration must be the true number of hydroxyl groups present for starting polyether chains. The desired initiating compound must be meticulously measured so that the number of hydroxyl groups is well known. Nonetheless, without a knowledge of the presence of water, the desired HEW will not be obtained.

From a review of Fig. 7, it can be said that the number of blocks in a copolymer depends on functionality. A surfactant molecule consisting of hydrophile-hydrophobe structure (alkylene oxides polymerized using a monol initiator) will behave differently than a hydrophobe-hydrophile-hydrophobe structure (oxide monomers polymerized with water or another diol) [77,96–101].

Molecular weight is also affected by the presence of water as an impurity. Size exclusion chromatography and other methods can be used to determine number average molecular weight (M_N), weight average molecular weight (M_W), and higher averages. Another method finding application in the determination of copolymer molecular weight is supercritical fluid chromatography [77,94]. While these methods give information about distributions, often this is not necessary. For many studies knowledge of the number average molecular weight is sufficient. Measurement of M_N is readily accomplished by wet chemical means. This is possible because of the relatively high concentration of hydroxyl groups in polyethers. Esterification of polyethers with organic acid anhydrides is an accepted industry method [102,103]. In this procedure, the acid moiety generated during esterification is titrated with standard base. The difference in milliequivalents of base needed to neutralize sample and blank is equal to the milliequivalents of OH in the sample. Knowledge of polymer functionality provides all that is necessary to find the number average molecular weight using this method.

When an analysis of a polyether of unknown functionality is performed, the number of OH groups per molecule, and therefore the number average molecular weight, cannot be determined. However, the hydroxyl equivalent weight can always be found because there is no requirement to know the functionality. It is common practice in the industry to assume the intended functionality and to use this number in converting HEW to molecular weight [104]. Those who do so without consideration of the underlying assumption are now forewarned.

TABLE 1 Impact of Water on HEW of Polyoxyethylene Made from MeOH. Ethylene Oxide Contains 200 ppm Water. Reaction Uses 32 g MeOH (1 Mole) and EO as Shown

HEW target	EO (g)	Total OH groups	HEW achieved	Average functionality	Mole fraction diol
10,000	9,968	1.222	8,190	1.100	9.97
20,000	19,968	1.444	13,850	1.182	18.20
40,000	39,968	1.888	21,180	1.308	30.80
80,000	79,968	2.777	28,810	1.470	47.00

Moles water = g EO * 200 ppm /18
Eq. OH from water = moles water/2
Moles MeOH = 1
HEW achieved = total g /total OH
Average functionality = total OH / total moles
Mole fraction diol = moles diol / total moles

Water is found in the alkylene oxide as well as the initiator mixture. Water from either source has the same impact on functionality. Continuous introduction of water with the oxide broadens the molecular weight distribution. Water in the oxide also puts an upper limit on M_N. Frequently too small to measure at low hydroxyl equivalent weight, the impact of water increases as higher HEW materials are prepared. Data in Table 1 illustrate the effects of water as attempts are made to reach hydroxyl equivalent weights in excess of 10,000. Note also that the average functionality increases as diol is mixed with monol due to the presence of water in the ethylene oxide. Table 1 contains calculated values based on 200 ppm water in the EO.

Examples of other contaminants which contribute OH functionality are seal fluids, cleaning solvents, and other polyalkylene oxide products from previous reactor use. These factors must not be overlooked in scale-up from laboratory to pilot scale to full production. At every step when the measured hydroxyl equivalent weight differs from the target, the presence of impurities in either initiator or alkylene oxides must not be ruled out.

During polymerization of ethylene oxide, side reactions are of minor consequence [63]; however, there is one troubling side reaction with EO, namely, the polymerization of EO during storage. High molecular weight polyethylene glycol (PEG) introduces haze in liquid PEG products [105]. PO and BO are not so well behaved in the polymerization process [30,106–108]. In addition to the desired nucleophilic reaction, alkoxide anions take part in an acid base reaction with the hydrogens *alpha* to the epoxide ring.* Two reports have stated that BO does not undergo the same rearrangement as PO [109,110]. Malhotra and Blanchard recognized unsaturation in the nuclear magnetic resonance spectra of anionically polymerized BO polymers prepared in their laboratory. The signal was attributed to a C_5 structure. Despite their inability to explain a mechanism to account for such a structure, the C_5 group was retained for the interpretation of their spectra [109]. Heatley et al. were unable to identify any unsaturation using ^{13}C NMR analysis [110]. It has been the experience of this laboratory to routinely find unsaturated products in many butylene oxide polymers having hydroxyl equivalent weights greater than 1,000 daltons [111,112]. While the conditions normally used in our work are more harsh than those used in the two reports described above, they are on a par with conditions used for PO polymerizations.

The rearrangement reaction and generation of new alkoxide anion initiator has been studied extensively for propylene oxide by the polyurethane industry [113]. The corresponding reaction with BO has not been discussed as often in the literature. Unpublished work in this laboratory has shown that isomerization of BO proceeds in the same fashion as PO [111]. BO and PO isomerization is typically controlled by reaction conditions [112]. Extraneous initiation during the course of polymerization introduces all of the same concerns of average functionality, molecular weight distributions and ultimate molecular weight as discussed previously in the case of water. More importantly to the preparation of block copolymers, allyl alcohol and crotyl alcohol are formed only while PO or BO are present. This can have a major affect on the final polymer. When the hydrophobic PO and BO monomers are reacted after the hydrophilic EO, isomerization and initiation result in homopolyethers containing no hydrophile whatsoever. When the hydrophobe is intended as a central block between hydrophilic blocks (triblock material), the monol initiator from isomerization will form diblock copolymer surfactant.

The factors determining the amount of isomerization are temperature and concentration of base. Increasing either of these increases the amount of extraneous initiation. Arrhenius plots indicate the reaction is second order (Fig. 12). Conversion of the carbanion to the unsaturated alkoxide ion is extremely fast.

Less intuitive factors influencing isomerization rates are polyether molecular weight and alkali metal counter ion. Rate of isomerization increases rapidly with hydroxyl equivalent weight [29,114]. The use of sodium counter ion results in greater rates than potassium counter ion, all other factors being equal [114].

---

* The isomerization products of PO and BO is a second source of extraneous initiation apart from water. While water in the alkaline oxide results in diol polyethers, the acid-base reaction of PO and BO with alkoxide results in monolic polyethers. Figure 12 the isomerization reactions of PO and BO.

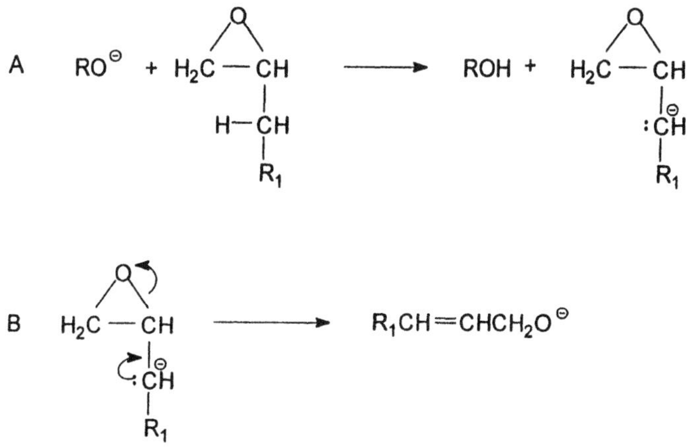

A $RO^{\ominus}$ + H_2C—CH(O)—CH(H)—R_1 ⟶ ROH + H_2C—CH(O)—$:CH^{\ominus}$—R_1

B H_2C—CH(O)—$:CH^{\ominus}$—R_1 ⟶ $R_1CH{=}CHCH_2O^{\ominus}$

FIG. 12 Isomerization reactions of PO and BO. (PO, R_1 = H; BO, R_1 = Me).

These effects can be attributed to differences in basicity of the alkoxide anion. Anions in solvent-separated ion pairs are better nucleophiles and weaker bases compared to anions in tight ion pairs. With the ion pair concept in mind, the dependency of isomerization on copolymer molecular weight and alkali metal counter ion is better understood. Sodium is the more tightly bound of the two metals commonly used. Increasing molecular weight decreases the concentration of hydroxyl groups needed to solvate and separate the ion pairs. A decrease in solvation leads to tight ion pairs which are less nucleophilic and more basic. The counter ion type has been shown to affect the rate of unsaturate formation. However, sodium and potassium alkoxides are identical in EO polymerizations [115].

Table 2 shows data relating isomerization to molecular weight in the polymerization of PO. As in Table 1, the values are calculated based on reasonable molar amounts of isomerization. Comparison of data in Tables 1 and 2 shows that isomerization affects molecular weight and functionality more than water in the oxide.

Extraneous initiation affects the polymer composition in many ways, including changes in surfactant performance. Average molecular weight, molecular weight distribution, the arrangement and size of polyoxyalkylene blocks, as well as the average functionality are each changed in some way by initiation during the course of the polymerization. Impurities which act as initiators at the beginning of the polymerization bring about changes in molecular weight, functionality, and, to a lesser extent, block size and structure. During the course of polymerization extra initiators arrive through impurities and isomerization of

TABLE 2 Impact of Propylene Oxide Isomerization on HEW. Polyoxypropylene Made with Propylene Glycol and PO Polymerization Initiated with 38 g of PG (1 Eq. of OH)

HEW target	Mole percent PO isomerization	Total OH groups	HEW achieved	Percent of target HEW achieved	Average functionality	Mole fraction monol
1,000	0.05	1.01	990	99	1.98	1.6
2,000	0.15	1.05	1,900	95	1.91	9.2
3,000	0.60	1.4	2,840	71	1.55	45
6,000	2.00	3.06	1,650	33	1.20	80

Percent isomerization varies with temperature and catalyst level.
Values used here are conservative.

PO and BO. In this way, molecular weight, molecular weight distribution, functionality, block size, and structure are all changed.

C. Other Common Synthetic Problems

This section is ambitious, as it is clearly impossible to list every problem associated with the preparation of a block copolymer. In laboratory, pilot-scale, and full-scale situations there are three problems which occur with enough regularity to deserve mention.

Changes made to reduce the polymerization cycle time cause the first problem. After a polyoxyalkylene block copolymer has been prepared and shown to be effective in an application, changes made in reactor conditions must be made carefully. As explained previously, preparation of block copolymers is carried out in a semibatch manner. Initiator and base are first charged to the reactor. Alkylene oxide is added at a rate appropriate for the heat removal capabilities of the system. Rates of addition vary from system to system, but in nearly every case a steady state concentration of unreacted monomer is desired. When the total amount of alkylene oxide has been delivered to the reactor, the oxide concentration will decay at a rate consistent with hydroxyl and base concentrations, reaction temperature, and the particular monomer involved. In the manufacture of block copolymers, it is desirable that levels of the first oxide be reduced to a very low level before the introduction of the next. Since reaction rates decrease in going from EO to PO to BO, the time required to react all of the monomer increases in the same order. Care must be exercised that the time period between oxide additions is sufficiently long such that a consistently low level of the first oxide is achieved. When a portion of the first oxide remains as the second is introduced, a mixed transition zone results between the two pure blocks. This is most likely to occur when EO follows BO. For example, changes in surfactancy could occur because the pseudotriblock arrangement (BO block–transition block–EO block) is not the same as the target diblock polyether (BO block–EO block).

Oxidation is the second common problem encountered during block copolymer synthesis. Neutral polyethers at ambient temperatures oxidize very slowly in air. Elevated temperatures and the presence of base greatly accelerate oxidation. Polyethers are prepared under such oxidation—accelerating conditions. Unless air is rigorously excluded, from the time initiator contacts base until the final polymer is neutral and cool, significant oxidation will occur. The products of oxidation are first peroxide, then aldehydes and ketones accompanied by chain cleavage. Finally, carboxylic acids, carbon dioxide, and other low molecular weight compounds are formed. Uncontrolled oxidation can render the polyether unsuitable for its surfactant use. After polymerization and neutralization, antioxidants can be effectively used to control oxidation [116].

The last common problem is encountered during the neutralization step. Without neutralization, hydroxide will be formed when the crude polyether encounters water. While ethers are not cleaved by base, other detrimental reactions occur such as aldol condensation, followed by elimination and the formation of color bodies [117]. Aqueous solutions prepared with unneutralized polyoxyalkylene polymers are themselves basic and can not be handled in the same fashion as neutral solutions. Over-neutralization with acid or failure to remove salts to a low, consistent value also changes the surfactant behavior from that desired.

IV. MODIFICATION OF POLYOXYALKYLENE BLOCK COPOLYMERS

As with any organic material, modification of block copolymers takes place by subjecting the material to suitable conditions and reagents. Modification may enhance or diminish the desired properties. Two broad types of modifying reactions are encountered: uncontrolled degradation or carefully manipulated procedures to impart new features.

A. Modification of the Hydroxyl Group

The synthetic utility of the OH group is well known to the organic chemist. This function is often the gateway to many diverse and interesting compounds [118]. March has enumerated in excess of one hundred reactions to prepare various alcohol derivatives [119]. With the possible conversion of these deriv-

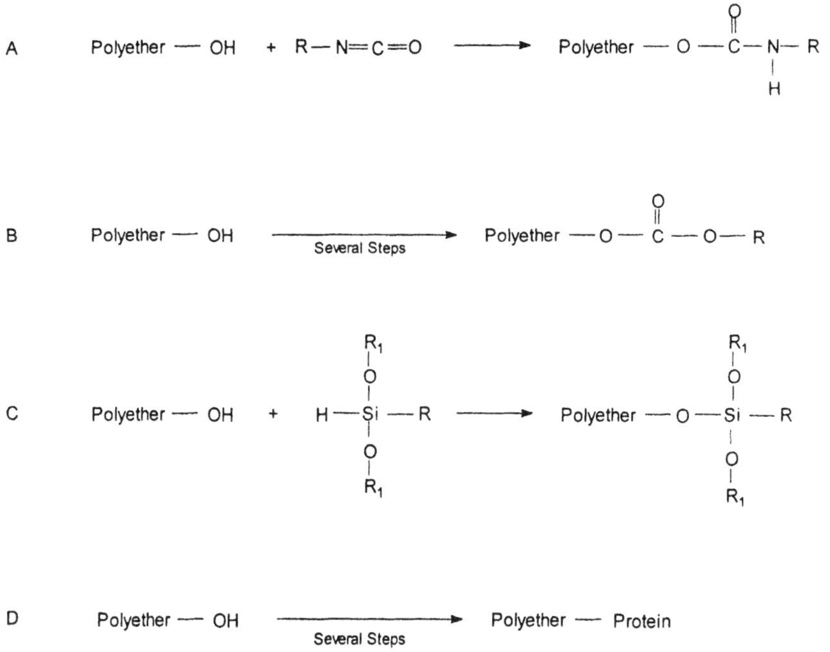

A Polyether — OH + R—N=C=O ⟶ Polyether — O — C(=O) — N(H) — R

B Polyether — OH ⟶ (Several Steps) Polyether — O — C(=O) — O — R

C Polyether — OH + H — Si(OR_1)(OR_1) — R ⟶ Polyether — O — Si(OR_1)(OR_1) — R

D Polyether — OH ⟶ (Several Steps) Polyether — Protein

FIG. 13 Example of polyoxyalkylene hydroxyl reactions in polyethers. A) Urethane. B) Gasoline additive. C) Graft silicone polyether R_1 represents silicone backbone. D) Pharmaceuticals.

atives to other functional groups, the potential for molecular modification through the OH group is tremendous. The terminal hydroxyl group in block copolyethers is no exception. It is because of this function that polyethers in general find application as chemical intermediates in diverse areas such as polyurethanes [121,122], gasoline additives [123], silicone-polyether surfactants [124–126], phase transfer catalysts [127], and pharmaceuticals [128,129]; amino group modifications of PEG materials have also been reported [130] (see Fig. 13 for typical reactions). The reader is also referred to the work by Bailey and Koleski for additional examples [131].

For surfactant applications, however, the terminal hydroxyl is most often left unchanged. It is sometimes useful to reduce or eliminate the polar interaction between water and terminal hydroxyl groups and keep other parameters, particularly block size and arrangement, constant. Polyoxyalkylene block copolymers which differ only in the terminal groups can be used to estimate the role surfactant polarity plays in overall performance. This is most easily done in the comparison of polyethers containing one, two, and zero hydroxyl groups. Figure

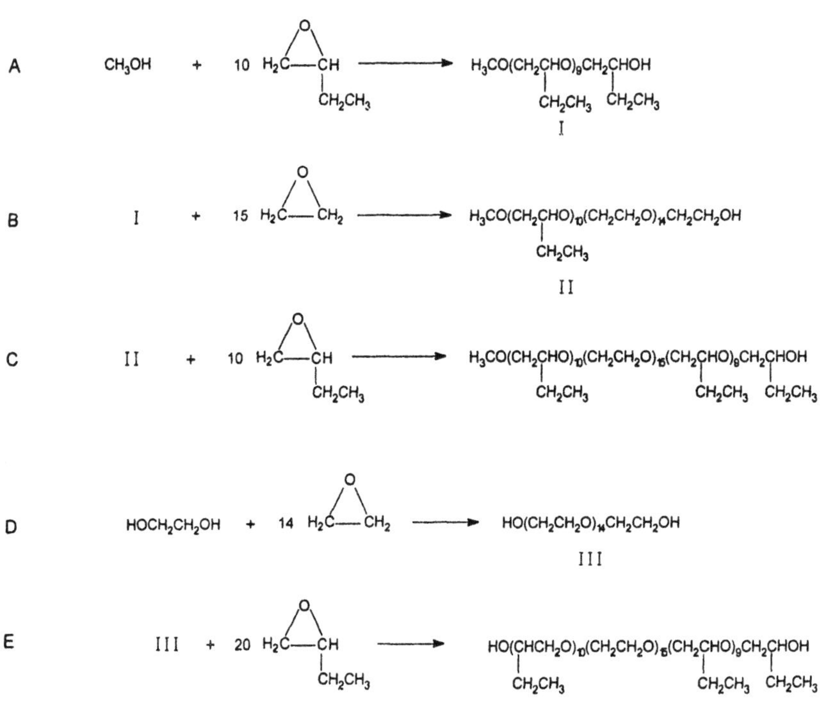

A $CH_3OH + 10\ H_2C\overset{O}{\text{—}}CH(CH_2CH_3) \longrightarrow H_3CO(CH_2\underset{CH_2CH_3}{C}HO)_9CH_2\underset{CH_2CH_3}{C}HOH$
I

B I $+ 15\ H_2C\overset{O}{\text{—}}CH_2 \longrightarrow H_3CO(CH_2\underset{CH_2CH_3}{C}HO)_{10}(CH_2CH_2O)_{14}CH_2CH_2OH$
II

C II $+ 10\ H_2C\overset{O}{\text{—}}CH(CH_2CH_3) \longrightarrow H_3CO(CH_2\underset{CH_2CH_3}{C}HO)_{10}(CH_2CH_2O)_{15}(CH_2\underset{CH_2CH_3}{C}HO)_9CH_2\underset{CH_2CH_3}{C}HOH$

D $HOCH_2CH_2OH + 14\ H_2C\overset{O}{\text{—}}CH_2 \longrightarrow HO(CH_2CH_2O)_{14}CH_2CH_2OH$
III

E III $+ 20\ H_2C\overset{O}{\text{—}}CH(CH_2CH_3) \longrightarrow HO(\underset{CH_2CH_3}{C}HCH_2O)_{10}(CH_2CH_2O)_{15}(CH_2\underset{CH_2CH_3}{C}HO)_9CH_2\underset{CH_2CH_3}{C}HOH$

FIG. 14 Scheme for preparing monols and diols with otherwise identical hydrophilic and hydrophobic blocks. Product of Reaction C is BO-EO-BO monol. Product of Reaction E is the identical BO-EO-BO copolymer, but is a diol.

14 shows a synthetic scheme for preparing block copolymers which differ only in the number of OH groups. Further reaction of the hydroxyl group on the product of Reaction C to give methyl ether would result in zero hydroxyl groups.

Thus, nearly identical polyoxyalkylene block copolymers containing two, one, or zero OH groups can be prepared. Extension of this idea to polymers having functionality greater than two is complicated by the number of polyether chains and arrangement of the hydrophilic and hydrophobic blocks.

The greatest barrier in using the method of hydroxyl group conversion is finding those conditions which give complete reaction [132–134]. Williamson-type etherification is one procedure where high conversion often requires complicated conditions [120]. The formation of esters or urethanes from carboxylic acid derivatives or isocyanates proceeds in high conversion under simple and generally mild conditions [135,136]. Many conversions have been done on the laboratory scale by first preparing tosylate or other sulfonate [137,138] and then reacting with other reagents. In these cases, tosylate conversions are high but final yields are low, making this scheme unattractive for industrial applications.

B. Modification of the Polyether Backbone

Reactions of the polyether backbone most often results in chain scission. As a consequence of this undesirable degradation the chain is seldom modified.

Oxidation is a frequently encountered chain cleaving reaction. Oxidation can be problem during reactions to modify terminal hydroxyl groups. Typically, oxidation is more of a problem with chain modification due to harsher conditions required. Oxidation proceeds by free radical processes [139,140].

Another method for modifying the copolyether chain would be to copolymerize other reactive monomers with EO, BO or PO; however, the comonomers would have to be carefully chosen.

While not truly a portion of the chain, the initiator is sometimes a site for modification. This is accomplished by using an initiator with functionality suitable for synthetic reactions but one which does not react with oxiranes under basic conditions [141,142].

REFERENCES

1. *The Polyglycol Handbook*, The Dow Chemical Company, Midland Michigan, 1988, pp. 13–28.
2. L. G. Lundsted and I. R. Schmolka, in *Block and Graft Copolymerization* Vol 2 (R. J. Ceresa, Ed.), John Wiley and Sons, New York, 1976, p. 13.
3. Ref. 2, pp. 8–9 and references therein.
4. N. Shachat and H. Greenwood, in *Nonionic Surfactants* (M. Schick, Ed.), Marcel Dekker, New York, 1967, p. 12.
5. N. Manolova, J. Libiszowski, R. Szymanski, and S. Penczek, *Polym. Int. 36*:23 (1995).
6. J. Kuyper and G. Boxhoorn, *J. Catal. 105*:163 (1987).
7. J. Dale, *Tetrahedron 49*:8707 (1993).
8. Z. Jedlinski, A. Dworak, and M. Bero, *Makromol. Chem. 180*:949 (1979).
9. F. M. Rabagliati and F. López, *Makromol. Chem. Rapid Commun. 6*:141 (1985).
10. T. Listoś, W. Kuran, and R. Siwiec, *J. Macromol. Sci. Pure Appl. Chem. A32*:393 (1995) and references therein.
11. T. Aida and S. Inoue, *Macromolecules 14*:1162 (1981).
12. J. E. McGraft, Y. Yoo, R. B. Turner, and D. M. Lewis, U.S. Patent 4,943,626 to The Dow Chemical Co., (1990).
13. Y. Yoo and J. E. McGraft, *Polymer (Korea) 18*:1063 (1994).

14. D. E. Laycock, U.S. Patent 4,962,281 to The Dow Chemical Company (1990).
15. F. Takrouri and K. Alyürük, *Polymer 35*:1518 (1994).
16. L. St. Pierre, in *High Polymers* Vol. 13 Part 1 (N. Gaylord, Ed.), Interscience, New York, 1962, p. 138.
17. A. Wurtz, *Compt. Rend. 49*:813 (1859).
18. A. Wurtz, *Ann. Chim. et Phys. 69*:317 (1863).
19. F. E. Bailey, Jr. and J. V. Koleske, *Alkylene Oxides and Their Polymers*, Marcel Dekker, New York, 1991, pp. 27–33 and references therein.
20. L. G. Lundsted, U.S. Patent 2,674,619 to Wyandotte Chemicals Corp. (1954).
21. E. Santacesaria, M. D. Sevio, L. Lisi, and D. Gelosa, *Ind. Eng. Chem. Res. 29*:719 (1990).
22. J. March, *Advanced Organic Chemistry*, 2nd Edition, McGraw-Hill, New York, 1977, pp. 340–341.
23. S. Inoue and T. Aida, in *Ring Opening Polymerization* Vol. 1 (K. J. Ivin and T. Saegusa, Eds.), Elsevier Applied Science, New York, 1984, p. 234.
24. A. Penati, C. Maffezzoni, and E. Moretti, *J. Appl. Polym. Sci. 26*:1059 (1981).
25. J. Lee, J. Kopecek, and J. Andrade, *Polym. Mater. Sci. Eng. 57*:613 (1987).
26. V. M. Nace, R. H. Whitmarsh, and M. W. Edens, *J. Am. Oil Chem. Soc. 71*:777 (1994).
27. P. L. Matlock and N. A. Clinton, *Chem. Ind. 48*:101 (1993).
28. K. J. Saunders, *Organic Polymer Chemistry*, Chapman and Hall, London, 1973, pp. 162–164.
29. D. M. Simons and J. J. Verbanc, *J. Polym. Sci. 44*:303 (1960).
30. J. Ding, F. Heatley, C. Price, and C. Booth, *Eur. Polym. J. 27*:895 (1991).
31. E. Santacesaria, M. Di Serio, R. Radici, L. Cavalli, and R. Garaffa, *Riv. Ital. Sostanze Grasse 68*:261 (1991).
32. J. V. Karabinos and E. J. Quinn, *J. Am. Oil Chem. Soc. 33*:223 (1956).
33. P. J. Flory, *J. Am. Chem. Soc. 62*:1561 (1940).
34. G. J. Stockburger and J. D. Brandner, *J. Am. Oil Chem. Soc. 40*:590 (1963).
35. P. R. Geissler, *J. Am. Oil Chem. Soc. 66*:685 (1989).
36. A. E. Johnson, Jr., P. R. Geissler, and L. D. Talley, *J. Am. Oil Chem. Soc. 67*:123 (1990).
37. L. Gold, *J. Chem. Phys. 20*:1651 (1952).
38. B. Weibull and B. Nycander, *Acta Chem. Scand. 8*:847 (1954).
39. L. Gold, *J. Chem. Phys. 28*:91 (1958).
40. Ref. 4, pp. 13–26.
41. L. T. Monson and W. J. Dickson. U. S. Patent 2,766,292 to Petrolite Corporation (1956).

42. S. D. Holland, U. S. Patent 3,053,903 to Shell Oil Company (1962).
43. R. M. Laird and R. E. Parker, *J. Chem. Soc. (B)* 1062 (1969).
44. P. Sallay, J. Morgós, L. Farkas, I. Rusznák, G. Veress, and B. Bartha, *J. Am. Oil Chem. Soc. 62*:824 (1985).
45. E. G. Lovett, *J. Org. Chem. 56*:2755 (1991).
46. G. Natta and E. Mantica, *J. Am. Chem. Soc. 74*:3152 (1952).
47. J. Pluciński, H. Prystasz, and F. Pawul, *Chemia Stosowana 25*:195 (1981).
48. K. S. Kazanskii and N. V. Ptitsyna, *Makromol. Chem. 190*:255 (1989).
49. W. B. Satkowski, S. K. Huang, and R. L. Liss, in *Nonionic Surfactants* (M. J. Schick, Ed.), Marcel Dekker, New York, 1967, p. 94.
50. A. D. Winquist, Jr. and L. F. Theiling, Jr., U.S. Patent 3,317,508 to Union Carbide Corporation (1967).
51. K. S. Kazanskii and N. V. Ptitsyna, *Vosokomol. Soedin. Ser. B 29*:351 (1987).
52. N. Garti, V. R. Kaufman, and A. Aserin, in *Nonionic Surfactants. Chemical Analysis* (J. Cross, Ed.), Marcel Dekker, New York, 1987, pp. 245–246.
53. J. Chlebicki and A. Zwiefka, *Tenside Surf. Deterg. 27*:266 (1990).
54. R. G. Duranleau, M. J. Plishka, and M. Cuscurida, U.S. Patent 5,182,025 to Texaco Chemical Co. (1993).
55. W. C. Ziegenhain, R. T. Jackson, and L. Rose, U.S. Patent 4,254,287 to Conoco Inc. (1981).
56. P. Novak, M. Ilavsky, P. Spasek, K. Bouchal, and P. Pavlas, *Angew. Makromol. Chem. 179*:87 (1990).
57. H. Schott, A. E. Royce, and S. K. Han, *J. Colloid Interface Sci. 98*:196 (1984).
58. L. Marszall, *Langmuir 6*:347 (1990).
59. M. J. Schick, *J. Colloid Sci. 17*:801 (1962).
60. D. L. Smith, *J. Am. Oil Chem. Soc. 68*:629 (1991).
61. I. R. Schmolka and A. J. Raymond, *J. Am. Oil Chem. Soc. 42*:1088 (1965).
62. R. Varadaraj, J. Bock, S. Zushma, N. Brons, and T. Colletti, *J. Colloid Interface Sci. 147*:387 (1991).
63. C. C. Price and D. D. Carmelite, *J. Am. Chem. Soc. 88*:4039 (1966).
64. B. Weibull, *Acta Chem. Scand. 49*:207 (1995).
65. K. Yang, U.S. Patent 4,239,917 to Conoco Inc. (1980).
66. Y. Gotoh, *Kobunshi Ronbunshu 51*:558 (1994).
67. R. H. Carr, J. Hernalsteen, and J. Devos, *J. Appl. Polym. Sci. 52*:1015 (1994).
68. M. C. Allen and D. E. Linder, *J. Am. Oil Chem. Soc. 58*:950 (1981).
69. N. Márquez, R. E. Antón, A. Usubillaga, and J. L. Salager, *J. Liq. Chromatogr. 17*:1147 (1994).

70. K. Rissler, U. Fuchslueger, and H. J. Grether, *J. Liq. Chromatogr. 17*:3109 (1994).
71. O. Güven, *Brit. Polym. J. 18*:391 (1986).
72. M. T. Belay and C. F. Poole, *J. Planar Chromatogr. 4*:424 (1991).
73. A. H. Silver and H. T. Kalinoski, *J. Am. Oil Chem. Soc. 69*:599 (1992).
74. H. M. Hagen, K. E. Landmark, and T. Greibrokk, *J. Microcol. Sep. 3*:27 (1991).
75. U. Just, H. Holzbauer, and M. Resch, *J. Chromatogr. A 667*:354 (1994).
76. A. K. Syahabana, M. Arime, and S. Satou, *Kanzei Chuo Burnsekishoho 33*:117 (1994).
77. J. M. Dust, Z. Fang, and J. M. Harris, *Macromolecules 23*:3742 (1990).
78. I. R. Schmolka, *J. Am. Oil Chem. Soc. 54*:110 (1977).
79. R. E. A. Escott and N. Mortimer, *J. Chromatogr. 553*:423 (1991).
80. A. Saraiva, O. Persson, and A. Fredenslund, *Fluid Phase Equil. 91*:291 (1993) and references therein.
81. L. Farkas, J. Morgós, P. Sallay, I. Rusznák, B. Bartha, and G. Veress, *J. Am. Oil Chem. Soc. 58*:650 (1981).
82. A. Kastens, in *High Polymers* Vol. 13 Part 1 (N. Gaylord, Ed.), Interscience, New York, 1962, p. 190.
83. G. A. Gladkovskii, L. P. Golovina, and E. V. Ryzhenkova, *Khim. Tekhnol. Vspenen. Plastmass*, 44 (1970).
84. P. Sallay, L. Farkas, and I. Rusznák, *J. Am. Oil Chem. Soc. 67*:209 (1990).
85. P. R. Geissler, A. E. Johnson, Jr., and L. D. Talley, *J. Am. Oil Chem. Soc. 68*:544 (1991).
86. F. Heatley, G. Yu, C. Booth, and T. G. Blease, *Eur. Polym. J. 27*:573 (1991).
87. S. Dickson, G. Yu, F. Heatley, and C. Booth, *Eur. Polym. J. 29*:281 (1993).
88. D. R. Weimer and D. E. Cooper, *J. Am. Oil Chem. Soc. 43*:440 (1966).
89. B. Trathnigg, D. Thamer, X. Yan, and S. Kinugasa, *J. Liq. Chromatogr. 16*:2439 (1993).
90. J. Cross, in *Nonionic Surfactants: Chemical Analysis* (J. Cross, Ed.), Marcel Dekker, New York, 1987, p. 22.
91. Ref. 16, p. 106.
92. Ref. 19, pp. 48–51.
93. M. Rösch, in *Nonionic Surfactants* (M. J. Schick, Ed.), Marcel Dekker, New York, 1967, pp. 783–785.
94. M. Daneshvar and E. Gulari, *J. Supercritical Fluids 5*:143 (1992).
95. N. Buschmann, *Comun. J. Com. Esp. Deterg. 25*:333 (1994).
96. J. M. Harris, *JMS Rev. Macromol. Chem. Phys. C25*:325 (1985).
97. R. De Vos and E. J. Goethals, *Polymer Bull. 15*:547 (1986).
98. Z. Zhou and B. Chu, *Macromolecules 20*:3089 (1987).

99. Z. Zhou and B. Chu, *Macromolecules 21*:2548 (1988).
100. F. M. Veronese, P. Caliceti, A. Pastorino, O. Schiavon, L. Sartore, L. Banci, and L. M. Scolaro, *J. Controlled Release 10*:145 (1989).
101. P. Linse, *Macromolecules 27*:2685 (1994).
102. Ref. 90, pp. 372–375.
103. ASTM D 4274–88, *American Society for Testing and Materials, Philadelphia*, 1993.
104. D. J. Sparrow and D. Thrope, in *Telechelic Polymers: Synthesis and Applications* (E. J. Goethals, Ed.), CRC Press, Boca Raton, Florida, 1989, p. 201.
105. T. H. Baize, *Ind. Eng. Chem. 53*:903 (1961).
106. G. Gee, W. C. E. Higginson, K. J. Taylor, and M. W. Trenholme, *J. Chem. Soc.* 4298 (1961).
107. E. C. Steiner, R. R. Pelletier, and R. O. Trucks, *J. Am. Chem. Soc. 86*:4678 (1964).
108. G. Yu, A. J. Masters, F. Heatley, C. Booth, and T. G. Blease, *Macromol. Chem. Phys. 195*:1517 (1994).
109. S. L. Malhotra and L. P. Blanchard, *J. Macromol. Sci. Chem. A11*:1809 (1977).
110. F. Heatley, G. Yu, W. Sun, E. J. Pywell, R. H. Mobbs, and C. Booth, *Eur. Polym. J. 26*:583 (1990).
111. R. H. Whitmarsh, unpublished results (1990).
112. R. H. Whitmarsh, unpublished results (1993).
113. R. Herrington, L. Nafziger, K. Hock, and R. Moore, in *Flexible Polyurethane Foams* (R. Herrington and K. Hock, Eds.), Dow Plastics, Midland, MI, 1991, p. 2.7.
114. S. D. Gagon, in *Encyclopedia of Polymer Science* Vol. 6 (H. F. Mark, Ed.), John Wiley and Sons, New York, 1986, p. 282 and references therein.
115. M. Adal, P. Flodin, E. Gottberg-Klingskog, and K. Holmberg, *Tenside Surf. Deterg. 31*:9 (1994).
116. Ref. 90, p. 18.
117. M. M. Cook and R. A. Milulski, Am. Oil Chem. Soc. National Meeting, San Francisco, 1979.
118. R. T. Morrison and R. N. Boyd, in *Organic Chemistry* Fifth Edition, Allyn and Bacon, Boston, 1987, p. 665.
119. Ref. 22, pp. 1166–1191.
120. B. Abribat, Y. Le Bigot, and A. Gaset, *Synthetic Comm. 24*:1773 (1994).
121. R. Herrington and R. Turner, in *Flexible Polyurethane Foams* (R. Herrington and K. Hock, Eds.), Dow Plastics, Midland, MI, 1991, p. 3.1.
122. A. Nathan, D. Bolikal, N. Vyavahare, S. Zalipsky, and J. Kohn, *Macromolecules 25*:4476 (1992).

123. R. A. Lewis and L. R. Honnen, U.S. Patent 4,191,537 to Chevron Research Company (1980).
124. L. A. Haluska, U.S. Patent 2,846,458 to Dow Corning Corporation (1956).
125. R. M. Bryant and H. F. Stewart, *J. Cell. Plast.* *9*:1 (1973).
126. M. R. Porter, *Handbook of Surfactants*, Chapman and Hall, New York, 1991, pp. 204–205.
127. J. M. Harris and M. G. Case, *J. Org. Chem.* *48*:5390 (1983).
128. S. Zalipsky, C. Gilon, and A. Zilkha, *Eur. Polym. J.* *19*:1177 (1983).
129. S. Zalipsky and G. Barany, *J. Bioact. Compat. Polymers* *5*:227 (1990).
130. S. Cammas, Y. Nagasaki, and K. Kataoka, *Bioconjugate Chem.* *6*:226 (1995).
131. Ref. 19, pp. 115–130.
132. S. Zalipsky, C. Gilon, and A. Zilkha, *J. Macromol. Sci. Chem.* *A21*:839 (1984).
133. M. A. Smith and K. B. Wagner, *Polymer Prep.* *28*:264 (1987).
134. K. Jankova and J. Kops, *J. Appl. Polym. Sci.* *54*:1027 (1994).
135. D. N. Bhattacharyya, S. Krishnan, R. Y. Kelkar, and S. V. Chikale, *J. Am. Oil Chem. Soc.* *61*:1925 (1984).
136. C.-G. Gölander, S. Jönsson, T. Vladkova, P. Stenius, and J. C. Eriksson, *Colloids and Surfaces* *21*:149 (1986).
137. J. M. Harris, E. C. Struck, M. G. Case, M. S. Paley, M. Yalpani, J. M. Van Alstine, and D. E. Brooks, *J. Polym. Sci. Polym. Chem. Ed.* 22:341 (1984).
138. J. M. Harris, M. R. Sedaghat-Herati, P. J. Sather, D. E. Brooks, and T. M. Fyles, in *Poly(Ethylene Glycol) Chemistry: Biotechnical and Biomedical Applications* (J. M. Harris, ed), Plenum Press, New York, 1992, p. 375.
139. P. J. F. Griffiths, J. G. Hughes, and G. S. Park, *Eur. Polym. J.* *29*:437 (1993).
140. H. Kaczmarek, L. Å. Lindén, and J. F. Rabek, *J. Polym. Sci.: Part A: Polym. Chem.* *33*:879 (1995).
141. B. Grüning and G. Koerner, *Tenside Surf. Deterg.* *26*:312 (1989).
142. S. H. Harris, A. C. Good, R. Good, and S. Guo, *Modern Paint and Coatings* *84*:20 (1994).

2

Chemical Analysis of Polyoxyalkylene Block Copolymers

HENRY T. KALINOSKI Analytical Chemistry Section, Unilever Research U.S., Edgewater, New Jersey

I.	Introduction	32
II.	Molecular Structure	33
	A. General characterization	34
	B. Average molecular weight	49
	C. EO/PO ratio	51
	D. Unsaturation	52
III.	Mixture Composition	53
	A. Residual monomers and volatile by-products	53
	B. Peroxides	55
	C. Carbonyl content	56
	D. Antioxidant	56
	E. Poly(ethylene glycol)	58
	References	58

I. INTRODUCTION

This chapter describes various techniques and procedures that are regularly used to characterize polyoxyalkylene copolymer surfactants. The approaches described are used to obtain information on the chemical composition of the complex mixtures comprising this class of surfactants. This includes the molecular structure characterization of the polymer chains as well as the identification and quantitation of the residual reactants, by-products, and additives present in surfactant mixtures. All these components combine to yield the overall bulk properties of the surfactants and a full understanding of their contribution to mixture composition is essential to understanding the behavior of the surfactants in use.

This chapter was composed from information gathered from a number of sources and is intended to make surfactant users and analysts aware of the parameters affecting the chemistry of surfactant mixtures. Of particular value are previous volumes in the *Surfactant Science Series* on nonionic surfactants [1–3] and surfactant analysis [4]. Details are given on the tools and approaches available to evaluate these parameters. This chapter covers both traditional, wet chemical, and instrumental techniques, not all of which are required to characterize any one particular surfactant mixture. Somewhat more space is given to some of the instrumental techniques, due to the complex nature of the data produced and the limited exposure some readers may have had with the techniques. As with any measurement science effort, it is important to understand the use to which the analytical information will be applied. With this understanding, the appropriate procedures and measurements can be conducted to obtain the required information in the most effective manner. It is also important for the analytical scientist to remain aware of the nature and appearance of the analytical information, as the data can contain clues to properties or problems beyond the scope of the original analytical investigation.

An effort has been made to compile a useful bibliography of recent references on the analytical techniques and procedures specifically applied to polyoxyalkylene copolymer surfactants. Many of the references contain other references to earlier or more fundamental studies and developments. This chapter is by no means the definitive work covering all possible aspects of the use of these surfactants and the nuances of the analytical approaches required in each application. In many cases, the specific details of how experiments are conducted or the resulting data are interpreted are not included. These specifics are left to the surfactant or analytical professional facing a specific challenge. Further, this chapter is not for complete novices as it will not make the uninitiated into experts. One does not, however, need to be an analytical or surfactant expert to gain useful information or insight into the solution of a problem from this chapter. The information included is to be used in combination with training previously acquired and other details of the system being studied.

There are areas that are consciously excluded, including those covered in other chapters in this volume. These include some standard characterization tests and bulk physical property evaluations (color, cloud point, melting point, Karl Fischer water, etc.) which are described in detail elsewhere [4–7]. Also excluded are analysis for these surfactants in water and the environment [4,8,9], separation from other surfactants [10–13] and the use of analytical techniques to study other properties. This last category includes such areas as the use of nuclear magnetic resonance (NMR) spectroscopy to study self-diffusion [14], gel-formation [15] or molecular motion [16,17]; the combination of NMR and light scattering to understand aggregation [18]; other studies of aggregation/associative behavior using light scattering [19–24], fluorescence [25] or chromatography [26–29]; the combination of light scattering and ultrasonic absorption to determine physical and thermal properties [30]; use of differential scanning calorimetry (DSC) to probe thermodynamic properties [31,32]; or the use of gas chromatography to categorize surfactant polarity [33]. It is clear from this list that the techniques of measurement science extend into many areas of importance to the surfactant scientist or engineer. It would be a disservice to attempt to address each of these areas in addition to the basics of surfactant characterization in a single chapter. Therefore, the reader is directed to the references listed or other chapters in this volume for a more thorough treatment of any individual application area.

The intent of this chapter is to define the role of measurement science regarding polyoxyalkylene surfactant mixtures in the commercial and industrial product research, formulation, application, and development environments. Some effort is made to highlight recent advances in measurement science and conjecture on implications for the analysis of polyoxyalkylene copolymer surfactants. The work should alert academic investigators and students to the aspects of these materials that are not adequately understood, the complexity of the analytical challenges, and areas where fundamental research could improve techniques or procedures. Overall, an appreciation of what is required to characterize such complex materials should be gained. This allows more informed decisions to be made regarding the application of these surfactants, their analyses, and analysis of resulting commercial and industrial products.

II. MOLECULAR STRUCTURE

The characterization of any surfactant material requires that its identity be suggested by qualitative tests and confirmed through specific identity tests. Once the identity of a surfactant is established, further tests are conducted to understand the material's specific chemical composition. This may require separation of the surfactant from other surfactants or other components of commercial products. The separation of EO/PO surfactants from mixtures is described elsewhere [4,8–13].

Standard qualitative tests for the presence and identity of nonionic surfactants are generally applicable to ethylene oxide/propylene oxide (EO/PO) block copolymers. These include color reactions and spectrophotometric determinations relying on the complexation of large cations by the alkoxylate portion of the molecules. A large lipophilic anion completes the ion pair, which can be measured via absorbance of ultraviolet (UV) or visible (Vis) light. Development of color depends on molar concentration as well as EO/PO ratio for block copolymer surfactants. The techniques must be calibrated against external standards and are inherently low in accuracy. As absolute standards of every possible EO/PO block copolymer do not exist, arbitrary standards must be chosen, leading to the inaccuracies. Further, any number of alkoxylates will yield similar results to nonspecific color reactions. A thorough summary of the various qualitative tests for nonionic surfactants, and their applicability to EO/PO block copolymers, is given in volume 40 of this series [4].

A. General Characterization

1. NMR

Nuclear magnetic resonance (NMR) spectrometry has been highly developed as a primary structural characterization tool for organic molecules. The hydrogen (^{1}H) and carbon (^{13}C) nuclei are most useful for organic and surfactant chemists. Modern synthetic chemists use NMR data as the benchmark against which all other spectroscopic information is evaluated. This is due to the highly recognizable chemical shifts displayed by atoms in characteristic functional groups. These chemical shifts are, in large part, due solely to the local atomic environment around the individual atoms. Signal intensities are directly proportional to the number of atoms sharing that environment. The simple proportionality ratios lead to a ready quantitation of the number of atoms in each environment, without requiring known standards for calibration. As these conditions are usually unique for each atom in a system (or a fairly small number of nearly equivalent atoms when dealing with protons), most chemists view NMR spectra as infallible fingerprints of the structure of organic molecules. Full structural identification of organic molecules is often made based solely on interpretation of NMR data. Many laboratories involved in the characterization and use of surfactants are equipped with an NMR instrument, with a superconducting magnet, operating at 200–300 MHz (75.5 MHz for ^{13}C). These systems feature sophisticated datasystems and software to allow mathematical conversion of NMR frequency signals into interpretable spectra. The datasystems also contain instrument control software to allow a number of different NMR experiments to be performed readily.

This is, however, a gross oversimplification of reality. Detailed descriptions of NMR spectrometry and its use in surfactant characterization is beyond the scope of this chapter. Reviews of proton and carbon NMR of surfactant materials have been published [34–36], including one in a companion volume in this series [37]. Therefore, no effort will be made to describe the many physiochemical characterization studies that have been accomplished using NMR measurements. A few basic concepts of the technique will be described to allow an appreciation of the advantages and limitations of this powerful technique. Specific examples of the use of NMR for polyoxyalkylene block copolymer characterization will be described.

NMR is most appropriately applied in the structural characterization of highly purified single materials. Significant quantities of the material are usually required and very high solubility in a suitable solvent aids in acquiring high quality spectra. Solid materials, even if suspended in a liquid solvent, give no useful NMR signals on instruments designed for high resolution (solution-state) operation. The phenomenon giving rise to NMR spectra is a relatively weak nuclear interaction with a magnetic field, so the technique is not particularly amenable to low levels of analyte. NMR spectrometry is not a "trace" technique and any quantitation of materials in a sample present below 5 mole percent is suspect. This has been overcome, somewhat, by modern high field instruments (500 MHz for proton, 125.5 MHz for ^{13}C), but these are not common due to their high cost.

All of this is not to say that NMR has no role in the characterization of surfactant mixtures. One simply needs to understand the complexity of the technique and the data generated. The skills required to obtain and interpret useful NMR spectra can be acquired and enhanced through diligence and experience. It must be fully understood, however, that the information provided through NMR represents statistical characterization parameters and does not reflect the structure of a single molecule. As the EO/PO surfactants are quite complex mixtures, results from NMR characterization must be considered in the context of other physical, chemical, and spectroscopic data obtained. Other procedures, such as chromatography, must also be employed for extensive characterization of EO/PO block copolymers.

Methodology using proton NMR spectrometry in the characterization of polyoxyalkylene surfactants (for determination of EO/PO ratio) is described as "routine" [38–41]. This type of characterization, however, generally refers to EO/PO adducts of fatty alcohols and alkylphenols. In those instances, information from NMR is similar to that obtained for ethoxylated nonionic surfactants (i.e., type of adduct, EO and PO content, and some partial mixture characterization). Overall compositions are determined through comparison of integrated intensities for resonance signals due to the methyl protons and the methine groups of the PO units with those of the methylene units (both EO and PO).

Proton NMR can be used to determine ratio of primary to secondary hydroxyl groups in copolymers. Only secondary hydroxyl signals are found for PO-terminated molecules. Extensive signal overlap in proton spectra (covering only 10 ppm) limits the amount of structural information available for EO/PO surfactants. Proton NMR can not be used to determine sequence information or differentiate between random and block copolymers.

In contrast to proton NMR (based on the high-abundance ^{1}H nucleus), the low-abundance (1.1%) ^{13}C natural isotope of carbon exhibits the proper nuclear spin state and gives rise to NMR signals. This requires spectra to be acquired over very long periods of time on today's modern Fourier transform instruments to ensure that a significant signal (and therefore signal-to-noise ratio) is recorded. The ability to acquire ^{13}C NMR spectra increases the amount of information available regarding an EO/PO copolymer surfactant. EO/PO distribution along the copolymer chain can be determined along with information on primary/secondary hydroxyl ratio [39,42,43]. Some techniques rely on another NMR-active nucleus (fluorine, ^{19}F, 100% abundance) to obtain this information following surfactant derivatization [43,44]. Standard methods developed for urethane foam raw materials [44] can be applied in the determination of percent EO in an EO/PO copolymer surfactant. Again a fluorinated derivative is formed, but information on EO is obtained using ^{13}C NMR.

The ^{13}C NMR technique can also be used to differentiate PO-EO-PO blocks from EO-PO-EO blocks [45]. These efforts have been greatly enhanced through the development and use of the higher magnetic field strength instruments, yielding greater chemical shift resolution [45–48]. The high field instruments and advanced spectral interpretation capabilities are required for random copolymers [45,49–52], but these materials are not the focus of this volume.

Figure 1 shows a portion of the ^{13}C NMR spectra of two EO/PO block copolymer surfactant mixtures (Pluronic® L-101 and Pluronic F-68). Information on average molecular structures are available in the signals with chemical shifts at 70 ppm (EO-related) and at 73 and 75 ppm (PO-related). Signals at 77 ppm are from the $CDCl_3$ solvent and can be used to calibrate the chemical shift axis. For the low-EO L-101 (Fig. 1, top), the integrated peak area for the 70 ppm peak will be low relative to that of the 73 and 75 ppm signals. Just the opposite is found for the high-EO F-68, where the 73 and 75 ppm signals are quite small (Fig. 1, bottom). As this technique determines a ratio based on

---

FIG. 1 (Opposite) Portions of carbon NMR spectra of Pluronic EO/PO copolymer surfactants. Spectra of samples of L-101, top, and F-68, bottom, were obtained on a Varian XL300 NMR spectrometer from $CDCl_3$ solution. Insets show signals from methyl carbons about 17 ppm. Spectra were obtained by Arnold Jensen, Unilever Research U.S.

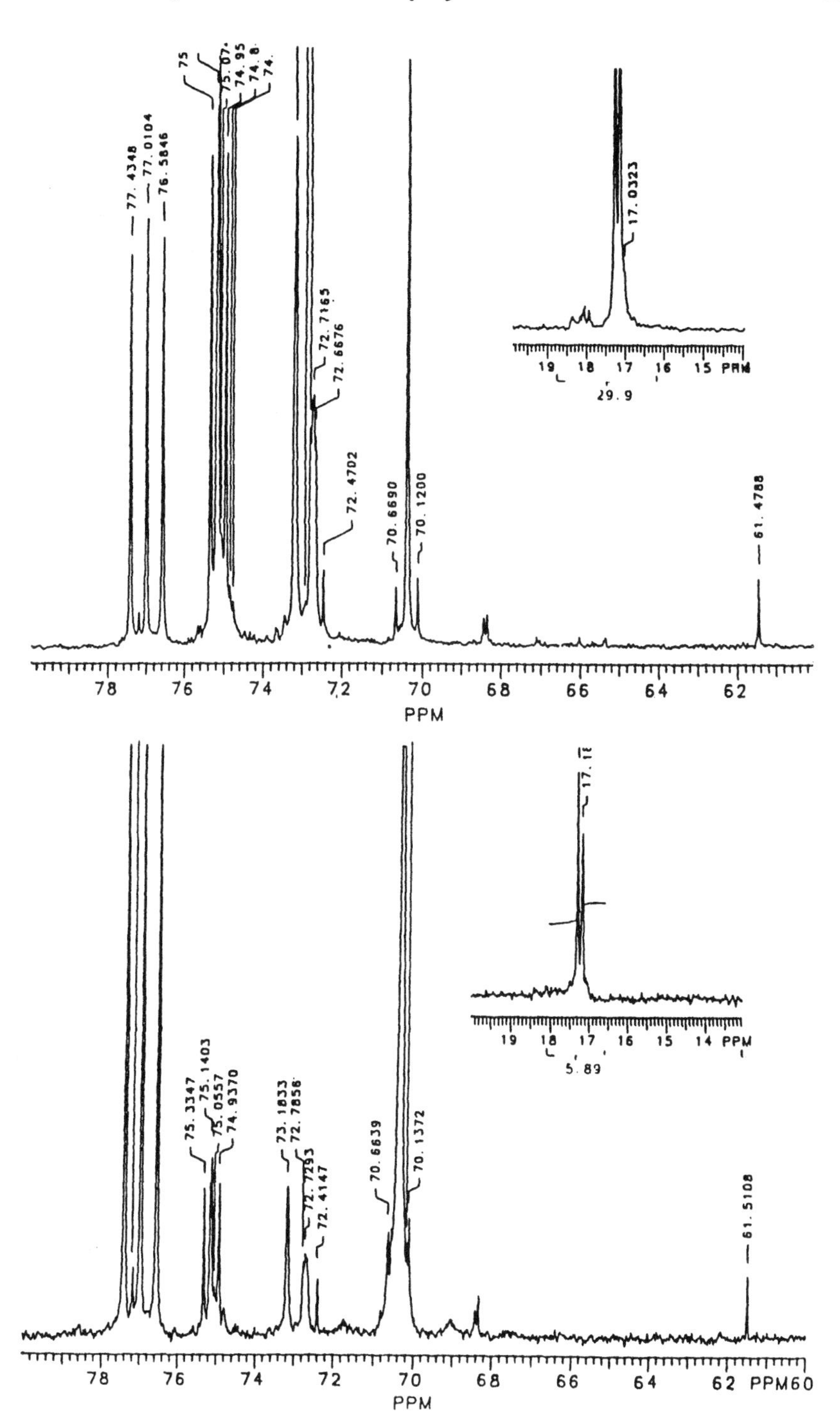
77.4348
77.0104
76.5846
75
75.07
74.95
74.8
74.
72.7165
72.6676
72.4702
70.6690
70.1200
61.4788
17.0323
29.9
PPM
75.3347
75.1403
75.0557
74.9370
73.1833
72.7856
72.7293
72.4147
70.6639
70.1372
61.5108
5.89
PPM

numbers of atoms, it is not essential to correct the calculation for molecular weight. In fact, NMR data can be used to determine average molecular weights for EO/PO copolymers [53]. Signals from the carbon in the terminal CH_2OH group are found around 61.5 ppm; methyl group signals appear around 17 ppm. Again, low-EO (high-PO; Fig. 1, top) samples can be distinguished from high-EO (low-PO; Fig. 1, bottom) materials by the integrated areas of these signals. More complete descriptions of chemical shift assignments and the use of NMR in the structural characterization of EO/PO block copolymer surfactants (both block and random) are available [47–50].

2. Vibrational Spectroscopy

Spectroscopic techniques probing the vibrations of bonded atoms in a molecule have long been a mainstay in chemical characterization laboratories. The systems generally use radiation in the infrared (IR) and near-infrared regions to excite molecular vibrations. Specific functional groups in organic molecules will absorb radiation (light) of a particular frequency. The loss of signal from the excitation radiation at a particular frequency is the indication of the presence of that specific functional group. The relative intensity of the signal loss is proportional to the overall abundance of the functional group in the sample. As the procedure operates simultaneously on the entire sample and all signals are summed and integrated, only average information on the structure and composition of a sample can be gained. This is generally not a problem when working with a single, highly purified material. Work on surfactant mixtures, however, must be considered with respect to information obtained through other techniques. Any number of general chemistry, analytical chemistry, and quantitative organic chemistry texts are available for detailed descriptions of the techniques.

Of interest to surfactant chemists are the developments leading to modern, computer-controlled spectrometers which have made it possible to obtain spectroscopic information on a wide variety of materials and mixtures. Improvements in hardware and software have led to the modern Fourier-transform (FT) IR spectrophotometer, which combines appropriate signal gathering and processing capabilities in a fast, sensitive, and highly accurate package. This is obtained through summing signals obtained through rapid accumulation of repetitive scans. Detector signals can be digitized for improved collection and manipulation. Scan-to-scan reproducibility is enhanced using computer control capabilities. Reviews of the technique applied to surfactant characterization have been published [4,54].

With appropriate accessories and sample cells, materials can be in any physical form, even as complex mixtures or small particles requiring IR-microscopy. Systems can be operated in reflectance as well as transmission modes, permitting characterization of surfactants deposited on surfaces. Reflectance techniques are somewhat less precise than transmission methods. The reflectance-type analyses are greatly enhanced through the use of spectral subtraction software, allowing background contributions to be removed. A technique called attenuated total reflectance (ATR) [55,56] allows even fully formulated surfactant-containing products to be analyzed directly using IR-spectroscopy. The approach does not require reproducible preparation of thin films and does not suffer from interference patterns found with liquid cells [57]. One well-developed application of the reflectance technique was the use of near-IR as a quality control procedure for EO/PO raw materials for urethane foams [58]. Although not directly a surfactant application, the approach used all the power of the spectroscopic technique with automatic, multiple wavelength measurements and computer-based data manipulation. A rapid method for hydroxyl number, water content and relative concentration of EO and PO was described. Again, however, a large number of well-characterized standard materials were required for external calibration.

A complementary, but distinct, vibrational spectroscopy technique is Raman spectroscopy [59–63]. In Raman, the frequency shift of incident radiation due to vibrational excitation of a molecule is detected. This is an inelastic scattering process which provides information if there is a change in the polarizability of a molecule due to the vibration. It has been observed that weak infrared absorbers are strong Raman scatterers, increasing the complementarity of the techniques. Further, water and silica are weak Raman scatterers and strong infrared absorbers. Raman spectra can be obtained under conditions unapproachable to IR (in/under water, in glass containers). (There are some limitations in aqueous systems for FT-Raman, which uses near-IR excitation radiation.) Weak Raman scatter by silica enables fiber-optic probes to be employed for remote sensing/sampling and in-situ applications.

Unfortunately, Raman signals are quite weak, with most of the scattering occurring at the incident frequency (elastic or Rayleigh scattering). Further, a great majority of materials fluoresce under conditions required for Raman spectroscopy. The fluorescence completely masks the Raman signal. This has been overcome, somewhat, by exciting samples at longer wavelengths, into the near-IR (1064 nm). Use of interferometers in Raman detection systems overcome the scattering loss deficiencies [64] of the longer wavelengths. The more efficient optical systems have also allowed use of lasers powerful enough to produce sufficient scattering but not so strong as to promote fluorescence. The interferometers also allowed for use of Fourier transform data manipulation in Raman spectroscopy [63,65,66].

Figures 2 and 3 show the IR and Raman spectra of two EO/PO copolymer surfactants, (Pluronic L-101 and Pluronic F-68). A comparison of the spectra clearly highlights the complementary nature of the techniques. Regions with strong IR absorbances (around 1100–1150 cm^{-1}) show weak Raman scatter. The strong Raman scatter around 2900 cm^{-1} is not quite matched by the IR

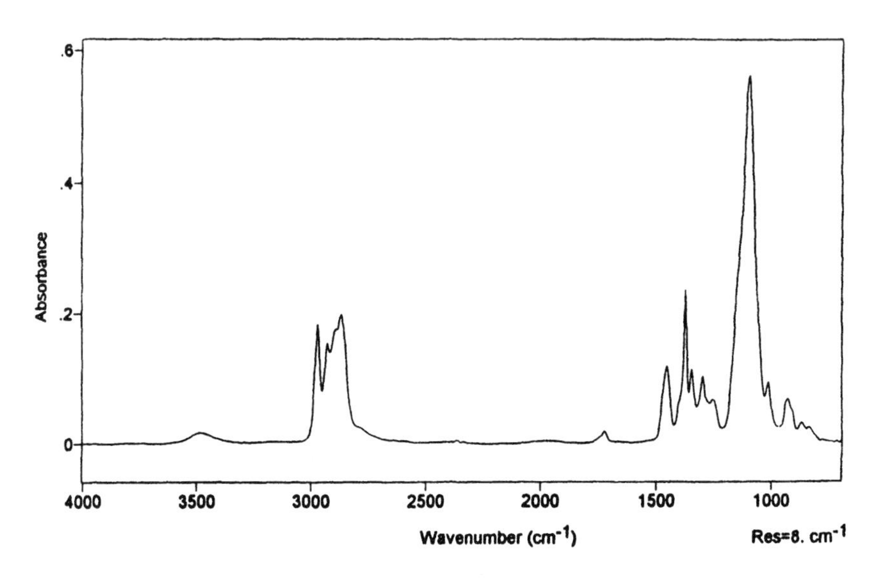

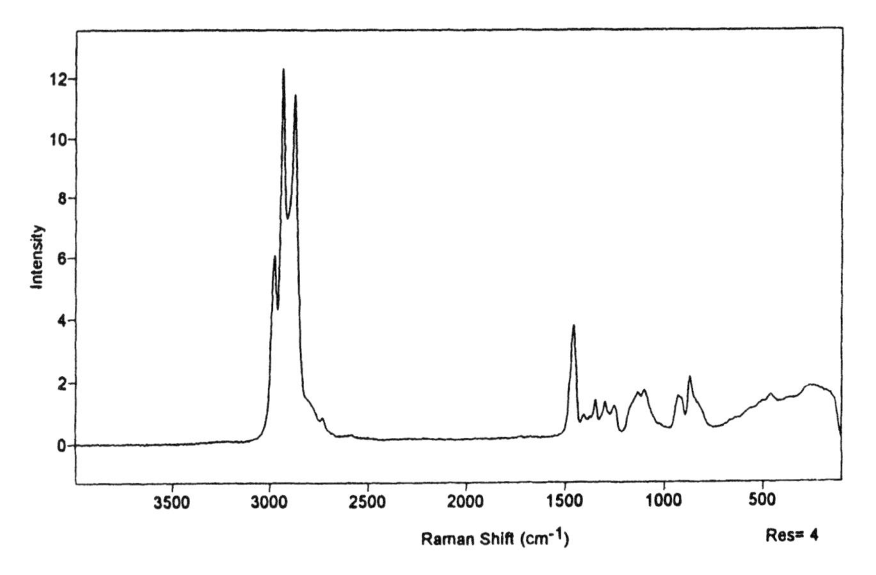

FIG. 2 FT-IR (top) and FT-Raman (bottom) spectra of Pluronic L-101 EO/PO copolymer surfactant. Spectra of the neat materials were obtained by Thomas Hancewicz, Unilever Research U.S.

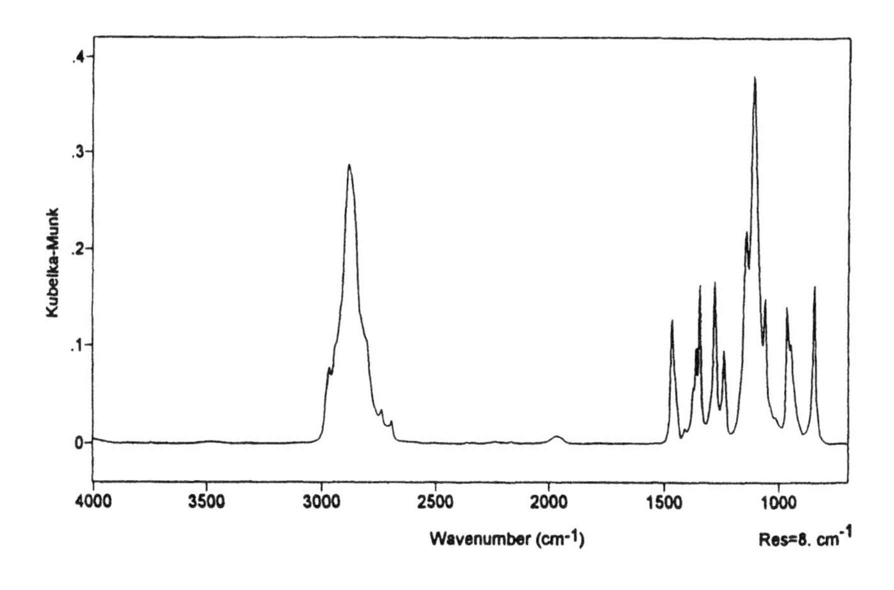

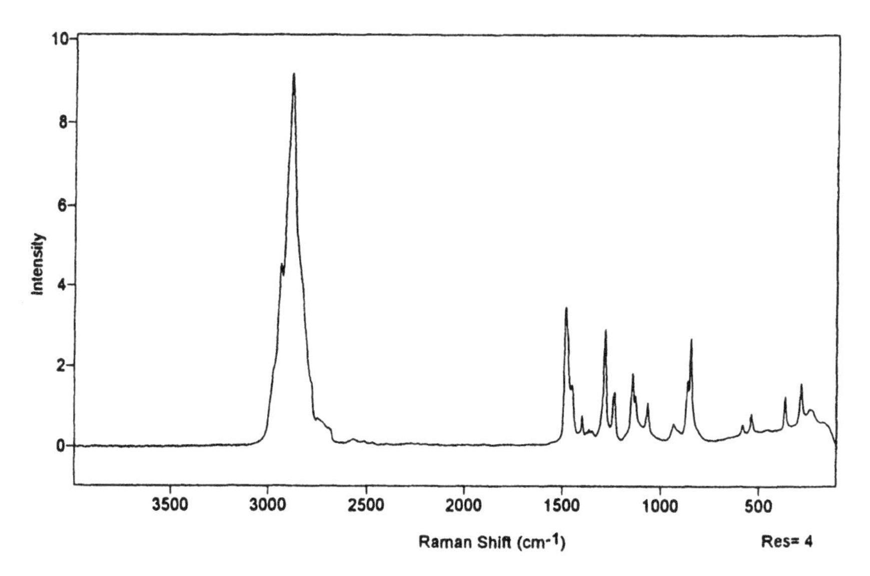

FIG. 3 FT-IR (top) and FT-Raman (bottom) spectra of Pluronic F-68 EO/PO copolymer surfactant. For FT-IR, the sample was ground with KBr and the spectrum obtained by diffuse reflectance. The FT-Raman spectrum was obtained on the neat solid. Spectra were obtained by Thomas Hancewicz, Unilever Research U.S.

spectra, particularly for the low-EO L-101 material (Fig. 2). In general, the 800–1600 cm^{-1} regions of the spectra contain information on the polymer backbone while the 2700–3000 cm^{-1} region reflects the pendant groups and chain ends. Specific assignments to either EO or PO groups can be made. More detailed assignments of IR absorbances or Raman scattering wavelengths are available [67].

It is interesting to note that the solid F-68 copolymer yields spectra characterized by sharp, well-defined absorbance bands (Fig. 3). The bands in the spectra of the liquid L-101 are far less distinct and poorly resolved. For both materials, this is more apparent in the Raman spectra. A sample of intermediate physical consistency (Pluronic P-104, a paste) exhibits bands of intermediate resolution. Also, note that spectroscopic information is available over a greater frequency range (below 500 cm^{-1}) in the Raman spectra than in the IR spectra. It appears likely with the increasing availability of high performance, computer-based FT-Raman spectrophotometers, this technique will play a larger role in the characterization of EO/PO surfactants. The quality of information available and the ease with which samples can be evaluated (virtually no sample preparation and straightforward coupling with fiber-optic probes) strongly support such future growth.

Infrared spectroscopy has recently been coupled directly with chromatography. Of significant interest to surfactant chemists are the interfaces with liquid chromatographic systems [68–70]. In one description of the technique [70], the composition of an ethylene-propylene copolymer was determined through comparison of C-H stretch absorbances due to methyl and methylene groups [71]. As the spectroscopy was coupled with chromatography, it could be seen how the composition varied as a function of molecular weight. The same type of information should also be readily available for EO/PO surfactants.

Specific application of vibrational spectroscopy to the characterization of EO/PO copolymer surfactants has been for both qualitative identification as well as quantitative analysis [72–74]. Vibrational spectroscopy can augment titration methods as a means of determining average molecular weight. Along with the derivatization and titration explained later, infrared spectroscopy can be used for determination of hydroxyl number [74–76]. As with other quantitative determinations using IR, hydroxyl number determinations require calibration against known, external standards. There are also matrix, hydrogen bonding, and water content complications to be considered. Appropriate choice of wavelength [75] or analysis from a solvent [77] can overcome some of these problems. Analysis in the near-IR can reduce difficulties presented by water absorbances due to the types of vibrations excited in the near-IR (generally only hydrogen) [76,78]. In the analysis of EO/PO copolymers, however, calibration curves must be prepared for each monomer ratio, as the calibrant chemical structure must match that of the analyte.

The EO and PO groups in copolymer surfactants give rise to a number of characteristic frequencies in the IR region. These absorbances (around 2300 cm^{-1} for $-CH_2-$, about 2485 cm^{-1} for oxygen in EO, 1380 cm^{-1} for PO methyl, and 1350 cm^{-1} for CH_2–O in PO) can be used to identify and characterize surfactant mixtures. The presence of a number of these absorbances is required to identify an unknown as an EO/PO copolymer since the absorbances are similar to those found for other ethoxylated or propoxylated surfactants.

3. Mass Spectrometry

A previous volume in this series contains an up-to-date summary of the state of this powerful analytical technique [79]; hence, the fundamentals of mass spectrometry (MS) will not be covered here. Application of mass spectrometry for the characterization of EO/PO surfactants has been fairly limited due to the relatively high average molecular weights of the copolymer mixtures. Previous efforts [80] concentrated on low molecular weight materials. Data on polymer composition, monomer distribution, and end-group analysis were gathered using electron ionization (EI) mass spectrometry. Even in this effort, however, the sample had to be derivatized to allow mass spectral information to be obtained.

Another approach to prepare polymeric materials for mass spectral analysis is pyrolysis [81,82] where the sample is heated until decomposition and the decomposition products identified using EI mass spectrometry. Pyrolysis can be coupled either directly to the mass spectrometer or first to gas chromatography (GC). The separation makes identification somewhat more straightforward and mass spectrometry as a detector for GC increases the information content available in a pyrogram. When applied to polyglycols, such as poly(ethylene glycol) and poly(propylene glycol), the products produced were terminal hydroxyl and alkene materials from cleavage of C-O bonds and water losses [81]. The polyether materials were found to be less stable to thermolysis than comparable polyolefins. This same approach could be applied to acquire some limited information on EO/PO copolymers.

The use of chemical ionization (CI) instead of EI for mass spectrometry can increase the amount of molecular weight information available for larger molecules. Figure 4 shows the ammonia positive ion CI mass spectra of two EO/PO copolymer mixtures (Pluronic L-101, top, and Pluronic F-68, bottom). In both cases, the average molecular weights of the samples are far in excess of what can be analyzed intact on the quadrupole mass spectrometer used for the analyses. The large majority of the ions found in both spectra are fragments of the general forms shown in Fig. 5 and are likely to arise from an in-source pyrolysis of the high molecular weight copolymers. The spectrum of the high-PO L-101 (Fig. 4, top) is characterized by ions predominantly composed of poly(propylene oxide) (R = CH_3, Fig. 5) and fragments are separated by 58 daltons. In contrast, the spectrum of the high-EO F-68 material is dominated by ions

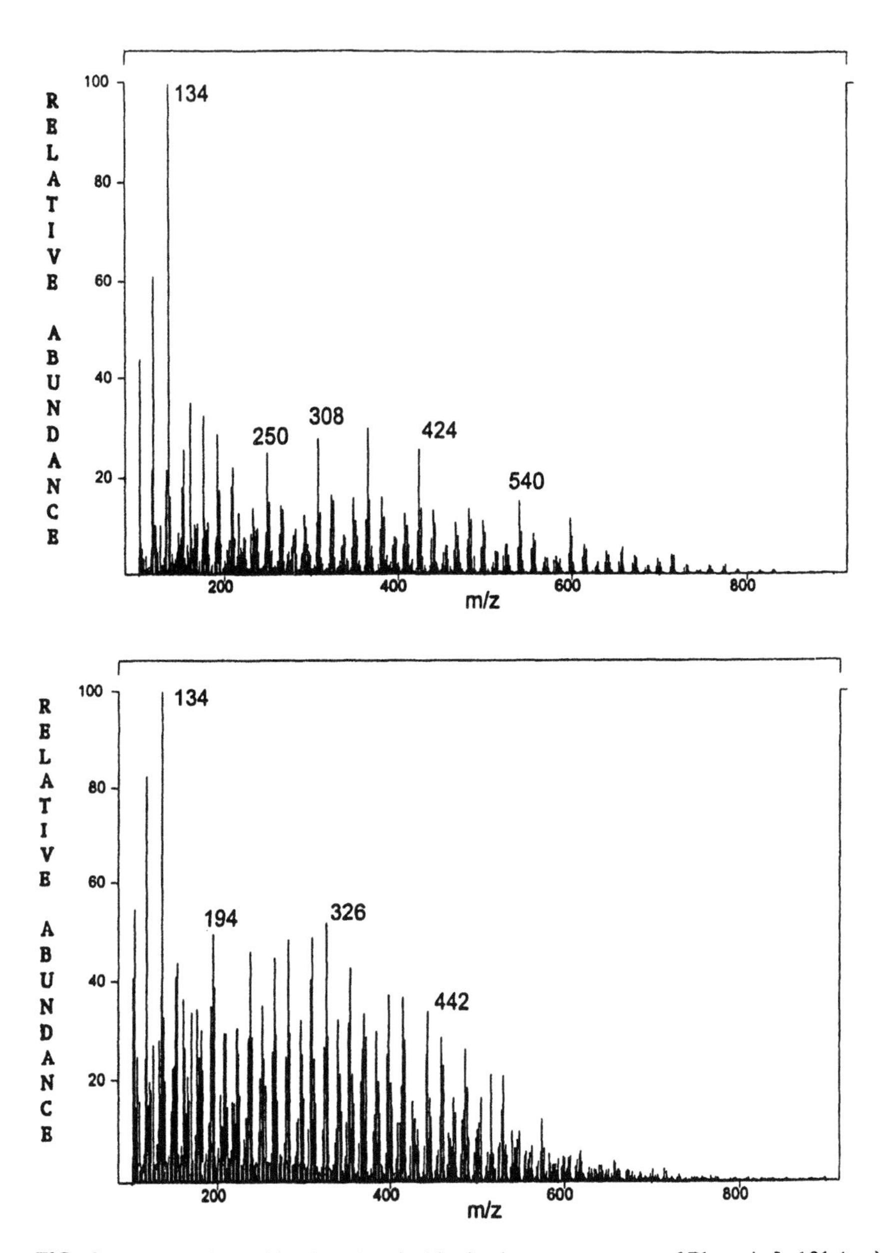

FIG. 4 Ammonia positive ion chemical ionization mass spectra of Pluronic L-101 (top) and F-68 (bottom) EO/PO copolymer surfactants. Spectra were obtained on a Finnigan-MAT SSQ710 single quadrupole mass spectrometer by Leonard Hargiss, Unilever Research U.S.

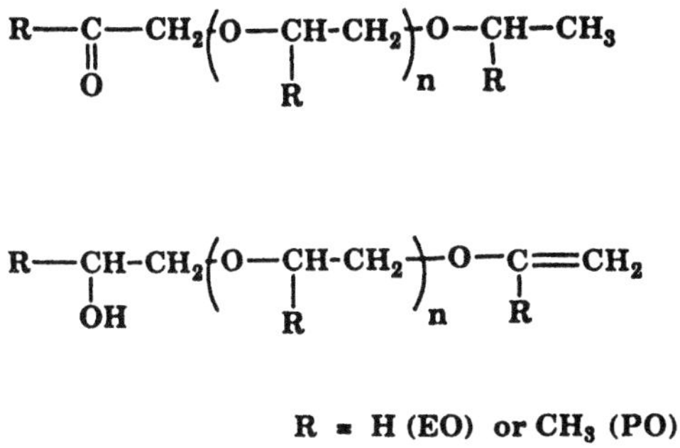

$$R-\underset{\substack{\|\\ O}}{C}-CH_2\left(O-\underset{\substack{|\\ R}}{CH}-CH_2\right)_n O-\underset{\substack{|\\ R}}{CH}-CH_3$$

$$R-\underset{\substack{|\\ OH}}{CH}-CH_2\left(O-\underset{\substack{|\\ R}}{CH}-CH_2\right)_n O-\underset{\substack{|\\ R}}{C}=CH_2$$

$$R = H\ (EO)\ \text{or}\ CH_3\ (PO)$$

FIG. 5 General structures of fragments formed in mass spectrometry of EO/PO copolymer surfactants. To form ions in mass spectrometry, charge can be supplied by protons, ammonium ions, or cations (e.g., sodium, lithium) in solution.

differing in molecular weight by 44 Daltons, primarily due to poly(ethylene oxide) fragments (R = H, Fig. 5). In both spectra, species are present as ammonium adduct ions $(M + NH_4)^+$, arising from the ammonia CI reagent. The ions found can be explained as arising from the same processes seen for pyrolysis [81] and in the collision-induced dissociation (CID) mass spectrometry of poly(alkylene glycols) [83,84] and alcohol ethoxylate surfactants [85,86]. These fragmentations are summarized in two very recent articles on the use of tandem mass spectrometry in the characterization of ethoxylated polymers [87,88]. Other ionization techniques (FAB and field desorption) appropriate for the analysis of low molecular weight EO/PO copolymers are described in these works. Included is an example of the characterization of an EO/PO copolymer (Jeffamine® M-360); however, this material is terminated with a butyl group and a primary amine functionality, which strongly direct the mass spectral fragmentation found.

Two recently developed techniques hold great promise for increased use of mass spectrometry in the characterization of EO/PO copolymers. Both the electrospray [89,90] and matrix-assisted laser desorption ionization (MALDI) techniques [91] have vastly increased the molecular weight range of analytes able to be addressed by mass spectrometry. Both techniques saw explosive growth and development based on the ability to address large biopolymers (i.e., peptides). Only now are the techniques being appreciated as tools for the characterization of synthetic polymers. Several reviews of both techniques have already been produced [92–96], so only details pertinent to the possible analysis of EO/PO copolymers will be given.

Electrospray ionization is a solution-phase technique in which a flow of somewhat polar solvent containing the analyte is subjected to a high electrical field [89,90,92,93]. There is still some controversy as to the actual mechanism of ion formation. Of note, however, is that the ionization is conducted at atmospheric pressure, not the high vacuum typical of electron, chemical, or other ionization techniques. This permits a wide range of samples and conditions to be addressed using electrospray. In fact, one of the first demonstrations of the technique by Prof. John Fenn was the ionization of a poly(ethylene glycol) sample with a number average molecular weight of 5,000,000 [92]. This was accomplished on an MS analyzer with only a few hundred Dalton mass range because electrospray has the ability to place several to several hundred charges on a single molecule. The mass-to-charge ratio (m/z) actually measured is still within the mass range of common quadrupole analyzers when z is much greater than one for very large molecules.

The use of electrospray for synthetic polymers is complicated by the fact that, unlike most peptides and proteins, polymers are complex mixtures of several hundred to several thousand oligomers. Even for low molecular weight polymers (several thousand Daltons), electrospray spectra are quite complex. This is further compounded when components of mixtures carry a large number of charges per molecule [97]. By adjusting the ion formation and transport conditions of an electrospray source, the spectra can be simplified. The process is further aided through use of data deconvolution software [97].

Figure 6 illustrates the appearance of electrospray mass spectra of EO/PO copolymer surfactants. Conditions used to obtain these spectra encouraged production of fragment ions and no ions related to the intact copolymers were found. The spectrum of the low-EO (high-PO) Pluronic L-101 (Fig. 6, top) is characterized by ions related to the predominant poly(propylene glycol) structure of the sample. Major ions differ in mass by 58 daltons. In contrast, the electrospray spectrum of the high-EO Pluronic F-68 (Fig. 6, bottom) shows primarily ions differing in mass by 44 daltons, related to the poly(ethylene glycol) structure. Structures of the ions in both cases are consistent with the ionization and decomposition mechanisms previously described.

Separations can be used to limit the complexity of mixtures entering the mass spectrometer for analysis. The electrospray ionization source has enabled a wide range of separation techniques to be coupled with mass spectrometry, including size exclusion (or gel-permeation) chromatography (SEC or GPC) [98,99]. The SEC/MS combination has proven effective as a tool to monitor copolymerization reactions by allowing composition and structural information to be correlated with molecular weight. The application of SEC/MS to the characterization of EO/PO copolymers is likely to be described in the near future.

Another approach to addressing spectral complexity is the use of analyzers with resolution sufficient to separate individual oligomers carrying multiple charges. Quadrupole analyzers and magnetic sector instruments do not possess this facility but ion cyclotron resonance (ICR) instruments do. These analyzers use the cyclic motion of ions in a strong magnetic field to resolve analyte

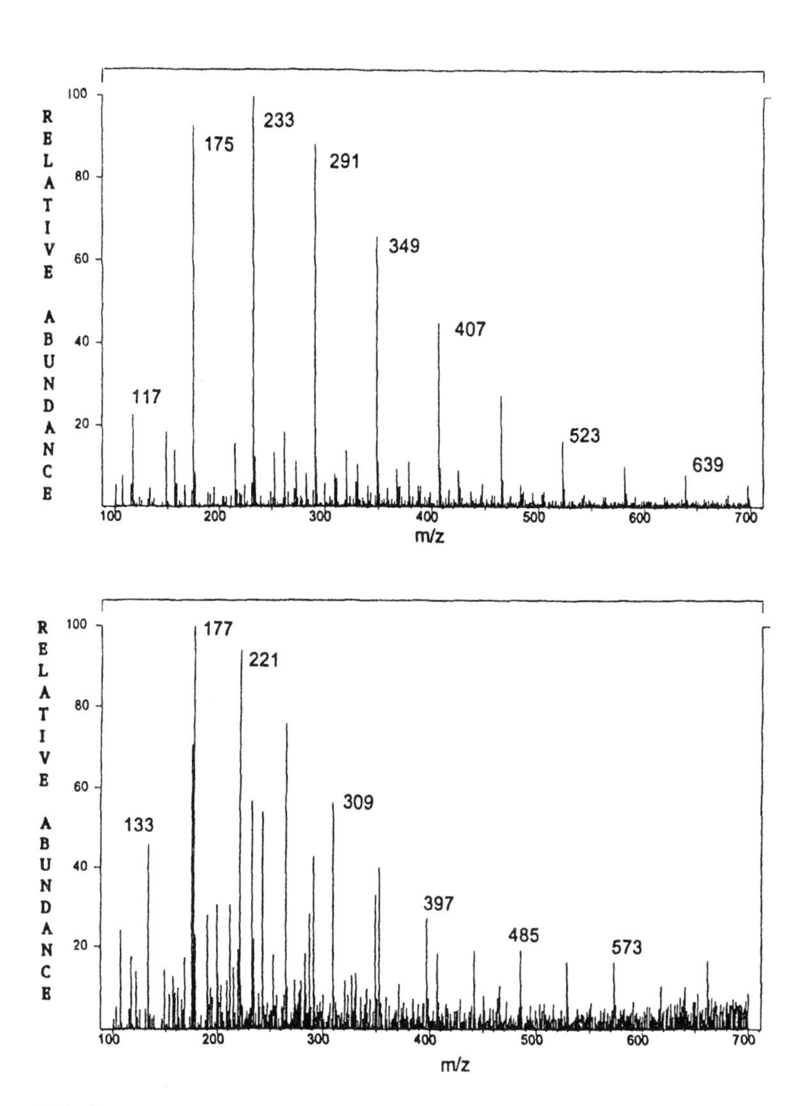

FIG. 6 Electrospray mass spectra of Pluronic L-101 (top) and F-68 (bottom) EO/PO copolymer surfactants. Spectra were obtained on a Finnigan-MAT TSQ70 triple quadrupole mass spectrometer using a Finnigan electrospray source. Samples were analyzed from a methanol:water solution. Spectra were obtained by Brian Shay of Unilever Research U.S.

species. The systems are also amenable to Fourier transform treatment of data, thereby increasing the flexibility of the analyses. Electrospray ionization has recently been coupled with FT-ICR instruments and applied in situations where high resolution has been required to solve identification problems [100]. One application described was the direct observation of multiply charged ions of poly(ethylene glycol) in the 21,000–23,000-Dalton mass range. Resolution of over 5,000 peaks, representing 65 oligomers in 12 charge states, was achieved [100,101]. Although ICR instruments are among the most expensive mass spectrometers, the ability to solve identification problems involving complex mixtures, such as EO/PO copolymer surfactants, is likely to make the equipment more common in industrial laboratories in the future.

The development of MALDI for mass spectrometry prompted the resurgence of one of the simplest mass spectrometer analyzers, the time-of-flight (TOF) instrument [102,103]. A laser light is used to desorb and ionize a sample that has previously been intimately mixed with an appropriate matrix. The matrix, usually an organic acid salt, plays a number of roles [104] primarily to isolate analyte molecules from one another and to absorb laser radiation. In the latter process, a large amount of energy can be deposited in the solid phase with a resulting disintegration of the solid. Little energy is transferred to the analyte molecules prior to this explosive process, limiting bond breakage in analyte molecules. As the firing of the laser for ionization can be precisely timed and controlled, the process is ideal for use with a TOF analyzer. Ion separation is accomplished based on the velocity of ions leaving the laser spot since flight time is directly proportional to ion mass. MALDI-TOF systems are among the simplest of mass spectrometers.

There are, of course, drawbacks to the system. There are uncertainties in the timing of the initiation event in MALDI-TOF. As flight-times are on the order of microseconds, a small uncertainty in establishing initiation leads to significant uncertainty in mass assignment. These systems are calibrated with compounds that are as close as possible, both chemically and structurally, to the analyte but calibrants do not behave identically with analytes. Not all molecules ionized in a single laser shot carry the same energy, so the energy spread of the moving ions translates into a distribution of flight times for ions of the same mass-to-charge ratio. Overall, time-of-flight is a low resolution mass spectral technique. This is further complicated when the sample is a polymer mixture, as opposed to a single peptide structure.

The most significant difficulty in MALDI is finding appropriate matrices, particularly for synthetic polymers. Most matrices identified in MALDI have been used for polar, water-soluble peptides and have not proven applicable to synthetic polymers. This is even true for poly(ethylene glycol) and poly(propylene glycol) samples [105,106]. Difficulties in protonating synthetic polymers and with samples not crystallizing properly have also been noted [105,106].

As with electrospray, the use of ICR analyzers with MALDI overcomes some of the limitations of the technique and improves the quality of data produced, such as in the analysis of poly(alkylene glycol) surfactants and polymers [107,108]. The coupling of MALDI with FT-ICR was found preferable over an SEC/MS combination using a particle-beam interface for the characterization of ethoxylate surfactants and polymer additives [109].

With the range of information available through MALDI (oligomer molecular weight, oligomer distribution, distribution symmetry, number and weight average molecular weights, monomer compositions, end group identification, etc.) and the amount of activity directed toward improving ionization and analysis conditions, it is very likely this technique will play a large role in future analyses of EO/PO copolymers.

B. Average Molecular Weight

1. Hydroxyl Number

A classical wet chemical method, using phthalic anhydride esterification [110,111], will yield information on the equivalent weight of an EO/PO block copolymer surfactant. A titration is used to determine the hydroxyl content of the analyte material. This value can be related to number average molecular weight but may differ from direct molecular weight determination due to the presence of impurities. This measurement is often more precise, however, than spectroscopic end group analysis as it does not require reference to a known molecular weight standard.

2. Chromatography

One type of liquid chromatography, size exclusion chromatography (SEC), utilizes the size of polymer molecules in solution to affect separation. As such, the behavior of a polymer mixture under controlled conditions and its elution from the separation column are related to the molecular weight of the polymer [112]. This combination offers a powerful approach to the determination of average molecular weight and molecular weight distribution for EO/PO copolymer surfactants. Unfortunately, the technique relies on size in solution and one does not know a priori the size of a polymer in solution. Further, the size of any polymer in solution depends on its composition, also not known prior to SEC evaluation. Determinations using SEC (also known as gel-permeation chromatography, GPC) generally utilize separation of known standards for assignment of molecular weights. Average molecular weight is determined through comparison of the elution volume of the analyte mixture to that of the standard. The ideal standards are materials of known composition and molecular weight that are identical to the test mixture. For EO/PO surfactants and copolymers in general, extremely well-characterized standards are not available. Often standards of poly(oxyethylene) or poly(oxypropylene) are used for EO/PO copolymers [28,29,113]. Separations can be conducted using either an aqueous or nonaqueous (usually tetrahydrofuran, THF) mobile phase. Aqueous mobile phases generally require high concentrations of salt modifiers to produce acceptable separations. As EO/PO copolymers have no UV absorbing chromophore, refractive index (RI) detection is generally employed. Peak shape and peak width are indicative of molecular weight distribution.

One thing that needs to be understood is that, unlike most polymers evaluated using SEC, EO/PO block copolymers are surfactants. This will lead to aggregation and micelle formation, significantly complicating attempts to use SEC for molecular weight characterization. To determine whether micelle formation is affecting SEC separation, the separation is conducted at a much different concentration of the analyte. If no concentration dependence is noted, it is unlikely that aggregate formation is occurring. The association behavior of surfactants [26] and EO/PO block copolymers [28,29,113] can be studied in detail using SEC.

Although most SEC molecular weight determinations rely on external standard calibrations, it is understood that polymers with equal elution volumes in SEC have equal hydrodynamic volumes in solution. Polymers elute in reverse order of their hydrodynamic volumes (i.e., large molecules elute first). The hydrodynamic volume of a polymer is simply related to its molecular weight and solution intrinsic viscosity. This premise, first developed by Benoit et al. [114], allows for the development of "universal calibration" procedures, independent of external standards. Specific detectors have been developed around this principle and are used to overcome the limitation of requiring known molecular weight standards [115–118]. This has significantly strengthened the capability of SEC for the characterization of copolymers.

As it is often desirable or necessary to use more than one detector to obtain sufficient information on a sample addressed by SEC, a number of efforts have been made to optimize multiple detector systems [119,120]. The advantage to such an approach is the ability to obtain all information required for complete characterization of a polymer mixture in a single chromatographic analysis. Common detector combinations include refractive index, viscometry, laser light scattering, and UV/Vis absorption. The detectors are coupled with a common data system to permit integration of information from all detectors. The UV detector would not be required for characterization of the EO/PO block copolymers.

As previously indicated, the coupling of SEC with spectroscopic detectors [68–70,98,99] significantly increases the amount of information available from a single chromatographic analysis. There are likely to be a great many applications of these techniques described in the near future.

Another approach that may aid in the characterization of EO/PO block copolymer surfactants is "critical condition" or "limiting condition" chromatography [121–123]. Not to be confused with supercritical fluid chromatography

(SFC), limiting condition separations are based on phenomena that allow retention volumes for a polymer to be independent of molar mass [121]. Under conditions of properly selected and controlled eluent composition, the polymer molecules are "invisible" to the size exclusion gel and the polymer elutes as though the gel is totally permeable. This is of advantage in the characterization of copolymers as the limiting conditions can be established for one portion of the molecule (rendering it invisible to the column) and have separation be based on the "visible" portion of the molecule. It must be noted that the invisible portion of the molecule must be a free end [124]. This procedure will not work at the limiting conditions of the central block of the triblock EO/PO copolymers. Limiting conditions have been established for poly(ethylene glycols) [125] and poly(propylene glycols) [126] and it remains to be seen if applications to EO/PO copolymers can be developed.

Another technique that may have promise in the characterization of EO/PO block copolymers is field-flow fractionation (FFF) [112]. This technique uses an externally applied field (e.g., gravitational, thermal, electrical, etc.) perpendicular to the direction of solvent flow in a thin channel to affect separation. As such, FFF is not a chromatographic technique. Although a few laboratories have investigated and published extensively on FFF, the technique does not appear to have caught on as a routine tool for industrial applications. Proponents claim the variations of the technique are more flexible and less problematic than SEC, and are applicable to virtually all polymers [127]. Further, as the thermal FFF subtechnique separates materials based on chemical composition as well as size, information on both size and composition of copolymers is obtained simultaneously [128]. It remains to be seen whether this technique will be applied in the characterization of EO/PO surfactants.

C. EO/PO Ratio

1. Vibrational Spectroscopy

To obtain EO/PO ratio information for EO/PO block copolymer surfactants, the instrument used must first be calibrated using known mixtures of polyethylene glycol (PEG) and polypropylene glycol (PPG). Calibration curves are prepared with ratios of characteristic absorbances plotted against known PEG or PPG percentages [40,72]. The near-infrared region and characteristic overtone bands can also be employed [72,78,129]. The IR techniques are accurate only over a limited range of EO/PO ratio.

2. NMR

As detailed in Section II, EO/PO ratio is directly available through comparison of integrated areas of signals arising from characteristic portions of the EO/PO molecules. Both proton and ^{13}C NMR can be used for these determinations. Proton NMR is generally faster than ^{13}C NMR and can be performed with less sample. Carbon NMR requires more spectral acquisition time, but spectra are better resolved and more easily interpreted. As there is no requirement for external calibration, NMR is favored for spectroscopic determination of EO/PO ratio.

3. Chromatography

Constituents of EO/PO surfactant mixtures are too involatile for direct analysis using gas chromatography. Even following derivatization, the molecular weights of these polymers preclude direct application of the technique. Sample preparation techniques have been developed, however, to gain useful information on EO/PO block copolymers using this versatile analytical tool. For the most part, however, these preparation-intensive cleavage reactions are not conducted if spectroscopic capabilities are available.

The EO/PO block copolymer surfactants can be cleaved through the reaction with hydrobromic acid (HBr) or hydroiodic acid (HI) to yield mixtures of bromoethane and bromopropane or iodoethane and iodopropane, respectively [130, 131]. The haloalkanes can then be separated and quantitated using GC. Some caution must be taken, as reaction yields for EO and PO differ, with lower amounts of halopropane formed [2,132–134]. EO/PO standards of known composition must be used to avoid erroneous results [2]. Acid catalysts can also be employed to increase reaction yield from PO [135]. The reaction with HI does not necessarily need to be followed with GC analysis, as the products of the reaction will lead to the production of titratible iodine [110,111,136]. As both EO and PO react to yield a mole of titratible iodine for each alkylene oxide unit, this approach will only allow determination of total alkylene oxide content.

A cleavage reaction using acetyl chloride and a ferric chloride catalyst is also applicable to the polyalkoxylate surfactants [137,138]. In this approach, chloroethyl and chloropropyl acetates are formed and an EO/PO ratio is determined by GC. The procedure is fairly straightforward to about 30% EO content with a calibration curve (based on known standards) required for higher levels [139].

Other cleavage reactions, using mixed acid anhydrides, are also applicable to EO/PO block copolymers [140–142] yielding glycol diacetates for GC analysis. Cleavage reactions can also be conducted in conjunction with analytical pyrolysis GC [82]. A cleavage reagent, in this instance phosphoric acid [143], is combined with the analyte mixture prior to pyrolysis. The reaction forms acetaldehyde and propionaldehyde for GC analysis. The procedure, however, relies on mixtures of poly(ethylene glycol) and poly(propylene glycol) of appropriate molecular weights for preparation of calibration curves.

D. Unsaturation

Chain termination reactions which occur during EO/PO block copolymer syntheses can lead to production of materials with terminal double bonds, instead of hydroxyl groups. Chains with terminal unsaturation are usually of lower molecular weight than the bulk of the EO/PO copolymer, resulting in lower average molecular weights for the polymer overall. Terminal unsaturations may also yield differences in surfactant physical and chemical properties. A number of techniques are available to determine unsaturation in an EO/PO copolymer mixture [144–147]. Spectroscopic methods (FT-IR, FT-Raman, and NMR) can indicate the presence of unsaturated materials but quantitation is better conducted using a specific chemical test. Titration of a mercuric acetate complex solution [146] is primarily used, being specific for terminal double bonds. One recommended adjustment to this procedure would be replacement of the specified carbon tetrachloride with another suitable solvent. The Wijs method for iodine number [147] can also be used if samples are scrupulously dry.

III. MIXTURE COMPOSITION

Aside from the primary polymeric components of an EO/PO block copolymer surfactant mixture, a sample will contain a variety of other materials. The materials present will depend on sample history and handling (i.e., manufacturing, storage, and use conditions) and will effect use and performance properties. Careful analysis of surfactant samples prior to use avoids extensive troubleshooting and detective work to determine causes of a failed product or formulation. The monomers and some by-products present safety and toxicology problems and their presence is regulated or prohibited. The nature and expected amount of the various mixture components are guides to the techniques which should be used in determinations. Some of the techniques provide information on more than one compound or compound type in a single analysis. It is often possible, for routine quality control, to analyze for only one or two suspected impurities, with high levels of one material indicative of the presence of others.

A. Residual Monomers and Volatile By-Products

The preparation of polyalkoxylated surfactants can give rise to a variety of small organic molecules in the final products. Although the manufacturing process usually includes vacuum-stripping at the end of alkoxylation, some residual monomer (ethylene or propylene oxides) or low molecular weight by-products (such as 1,4–dioxane, acetaldehyde, propionaldehyde, and allyl alcohol) may remain in the surfactant mixture. These materials are highly reactive and, as a result, toxic or carcinogenic [148–153].

A variety of methods have been developed to analyze for such materials in a range of ethoxylated surfactants and most are applicable in the analysis of EO/PO block copolymer surfactants. The analytical procedures used employ either gas or liquid chromatography (LC), with some form of sample pretreatment. Direct injection into a GC system, both with and without precolumns to remove much of the surfactant, was employed previously but was not highly reproducible and led to build-up of residual surfactant in the injection port, precolumn, and separation column. The only recourse when system performance ultimately degraded (usually after only a few injections) was to remove and clean or replace the separation column.

The most straightforward and commonly applied sample preparations use vacuum distillation [154,155] or headspace for GC [156–158] and solid-phase extraction for LC [159]. The solid-phase extraction techniques may prove applicable for GC analyses as well. Solvent extraction of 1,4–dioxane followed by GC has been used, but poor recovery and variability with matrix [160] has been found. Both vacuum distillation and headspace techniques are accurate and sensitive (part-per-billion detection limits are possible) but distillation is time-consuming and not amenable to automation.

Headspace sampling for GC relies on the partition of volatile materials present in a sample between the liquid and vapor phases in a sealed vial. The sample vial is often heated to increase the amount of material in the vapor phase. An aliquot of the headspace is removed for injection onto the GC column. This can be accomplished manually using a gas-tight syringe or automatically using specifically designed hardware. Automation significantly increases the number of samples and standards that can be addressed, and allows for minimum supervision of procedures. The sensitivity of the technique can be improved by operating in a dynamic mode, where the sample is continually swept with an inert gas, usually helium, and the volatiles collected on a sorbent trap. Again, heating increases the level of volatiles in the vapor phase. The sorbent trap is heated to release the collected vapors for injection into the GC. Commercial equipment is readily available for this approach as well. Dynamic headspace can be difficult for some materials as desorption from the trap is compound dependent. Procedures must be carefully developed for the compounds of interest.

Packed column GC often provides sufficient resolution for most samples. In cases where samples contain a number of low molecular weight volatile components, the resolution of capillary GC may be required. Capillary GC has the added advantage of lower detection limits.

As different surfactants have different affinities for these small organic materials, standards of the target analytes are prepared in the same or similar surfactant to that being analyzed. This ensures the same partition ratios for standards and samples. A means to improve accuracy and precision further is the use of mass spectrometric detection following GC [156]. A deuterated analog of the analyte is added directly to the surfactant as an internal standard (isotope dilution) and mass spectrometry differentiates between standard and sample based on distinct molecular weights. Standards of liquid materials (e.g., PO, acetaldehyde, dioxane) are straightforward to prepare while gaseous EO requires a more cautious, complex approach [158].

A new sampling device for GC, short path thermal desorption, has recently been developed and may provide advantages for simultaneous GC determination of volatile components in EO/PO surfactants. The approach permits ready interface with capillary GC and avoids the complication of injecting small quantities of gaseous headspace samples. Procedures and precautions in sample and standard handling for headspace methods would still apply.

Liquid chromatographic methods are used for selective detection of aldehydes in EO/PO surfactants. Aside from production during polymer synthesis, aldehydes are formed in EO/PO surfactants through oxidative decomposition (exposure to air and high temperatures). For LC, the aldehydes are derivatized using 2,4–dinitrophenylhydrazine (DNPH) with selective UV-absorbance detection [161]. Proper choice of UV wavelength (365 nm) makes detection specific for DNPH, avoiding impurities and increasing sensitivity. Reverse-phase LC on a siloxane-modified silica column is conducted soon after derivative preparation to avoid aldehyde-forming side reactions from the derivatization reagents. The derivatization conditions are somewhat harsh which may lead to acid-catalyzed hydrolysis of EO/PO surfactants and artificially high levels of acetaldehyde. Such hydrolysis does not occur in conditions used for GC.

B. Peroxides

Peroxides are formed in EO/PO block copolymers through oxidative degradation but are not the ultimate products from the reaction. Continued decomposition of the peroxides results in formation of aldehydes and acids. The process quickly reaches a steady-state equilibrium with maximum peroxide levels in the ppm range. Measurement for peroxide content gives an indication, but not a full picture, to the extent of mixture decomposition. A more complete evaluation of polymer decomposition is obtained through quantitation of aldehyde and other carbonyl compounds.

Levels of peroxides in EO/PO surfactants are determined through measurement of the oxidizing capacity of the sample. Iodometric procedures which reduce the peroxide compounds are generally employed [162,163]. These tests are not specific for any one peroxide and some peroxides are more readily reduced than others. As with any analytical measurement, the procedure used, titration [164] or spectrophotometry [165,166], should be specified. Some adjustments may be required to optimize conditions for EO/PO surfactants [166].

C. Carbonyl Content

EO/PO surfactants do not contain a carbonyl functionality as a part of their normal molecular structure. Oxidative decomposition, as mentioned previously, leads to formation of carbonyl compounds, usually aldehydes. Although small quantities of aldehydes may be formed during synthesis, measurement of carbonyl compounds is generally used as an indication of decomposition of EO/PO surfactants.

For the most part, an overall measurement of total carbonyl content is sufficient to indicate extent of surfactant degradation. As such, methods for carbonyl content are specific for the carbonyl functionality but not for any one compound. To quantitate levels of specific aldehydes, the LC methods for volatile components described above are recommended.

As a measure of gross contamination or surfactant degradation, vibrational spectroscopy is the rapid and accurate method of choice. The signal due to absorbance of infrared light in the 1700 cm^{-1} region is indicative of the carbonyl functionality. Samples can be examined either neat or in solution. The test is not sensitive (good to about 0.1%) so it is not applicable for trace contamination.

The DNPH derivatives of aldehydes prepared for LC will develop a red color on treatment with base. This color reaction forms the basis of the spectrophotometric measurement of total carbonyl content [167,168]. The test can be quite sensitive so care must be taken to exclude common carbonyl-containing compounds (e.g., acetone) from the area, glassware and solvent used for performing measurements.

Titration of evolved acid following reaction with carbonyl compounds can also be performed [169]. Corrections are made for inherent acidity or alkalinity of samples, as determined through blank titrations.

D. Antioxidant

Because EO/PO block copolymers are susceptible to oxidative decomposition, antioxidants are often added to commercial preparations to stabilize the materials. Usually a single material is added, substituted phenols being common. The aromatic phenol ring gives a readily accessible "handle" for characterization and UV spectrophotometry at the appropriate wavelength is sufficient for detection and quantitation. A range of separation techniques—GC, LC, or even SFC [170]—can also be employed for characterization. Mass spectrometry can be used for positive identification of the specific antioxidant employed. The presence of an antioxidant in a sample will interfere with determination of peroxide content. When the antioxidant is determined first, values obtained are used to correct the peroxide content value. If determination of peroxide level

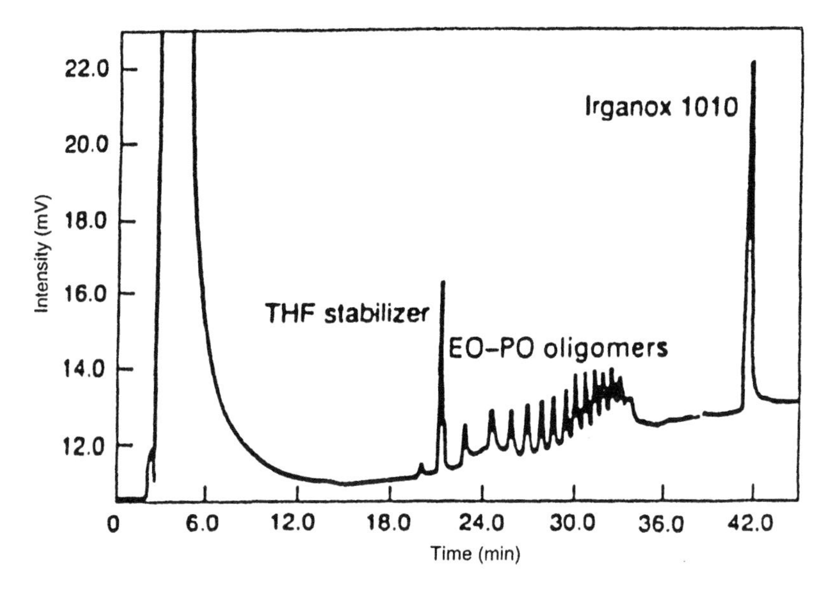

FIG. 7 Capillary SFC chromatogram of a poly(alkylene glycol) mixture extracted at 20 MPa with carbon dioxide, passed through a silica column. Details of extraction given in Reference 185. *Source*: from [185] with permission of the author and the Royal Society of Chemistry.

is attempted first and a negative result is obtained, the presence of an antioxidant is often indicated.

It might seem desirable to extract the antioxidant from the surfactant for a more definitive quantitation. This would normally be accomplished using an organic solvent to extract a basic solution of the surfactant. The emulsifying properties of these surfactants, however, make this a difficult task. This is an area where supercritical fluid extraction (SFE), using carbon dioxide as the solvent, may improve sample preparation [172–173]. The solubilizing power of the solvent can be tailored to the target antioxidant using a combination of temperature and pressure (and therefore, density), leaving much of the surfactant in the extraction cell. The process can be completed by collecting the extract off-line or interfaced directly with GC, LC, or SFC [174–184]. With a direct interface to chromatography, SFE may also prove effective for volatile, low molecular weight impurities. This procedure with off-line sample collection has recently been reported for antioxidants in EO/PO block copolymers [185]. This appears to be an ideal application for this newly emerging sample preparation technique. As shown in Fig. 7, the antioxidant Irganox 1010 was extracted using SFE and identified using capillary SFC, GPC, and NMR. The approach addressed liquid copolymer mixtures and relied on a silica adsorbent to retain much of the surfactant. Further work is still required to minimize coextraction of low molecular weight EO/PO oligomers.

E. Poly(Ethylene Glycol)

The presence of water during the synthesis of an EO/PO block copolymer can lead to production of by-product polyoxyalkylene. As ethylene oxide is usually added last for EO/PO block copolymers, the by-product is poly(ethylene glycol) (PEG). The average molecular weight of by-product PEG is generally lower than the principal surfactant. The presence of PEG will contribute to NMR signals when this procedure is used for characterization [186] leading to somewhat inaccurate determinations. Further, PEG is not a surfactant and will modify physical, chemical, and surfactant properties of the EO/PO surfactant. The chain termination reactions for PEG often lead to the formation of the terminal unsaturation. Measurement of unsaturation, described above, will unspecifically account for much of the PEG by-product.

Determination of PEG in EO/PO block copolymer surfactants is accomplished through separation of the PEG from the surfactant. This is facilitated by the usually lower molecular weight and different solubility of the PEG. Extraction of PEG from surfactants is possible [187–190] but often difficult due to the properties of the surfactant. Gas chromatography can characterize the PEG after isolation, with derivatization (trimethylsilyl or acetate) of the PEG required [190].

Liquid chromatography is the preferred method as little sample preparation is required and direct determinations can be made [189,191,192]. Quantitation is performed against a known standard of pure PEG of appropriate molecular weight. A molecular weight determination on the PEG can also be accomplished [191]. Using these reversed-phase procedures, PEG exhibits relatively short retention times. To avoid interferences from the "water dip" and unretained materials, a size-exclusion column can be coupled to the analytical LC column [192].

Supercritical fluid chromatography may prove effective in the determination of PEG in EO/PO block copolymer surfactants [170,186,193,194]. PEG is readily soluble in supercritical carbon dioxide to a relatively high molecular weight. This can be extended through normal derivatization procedures. SFC conditions, on either packed or capillary columns, may be able to be developed to allow characterization of only the PEG or possibly the PEG and surfactant both. This is an area deserving additional attention.

REFERENCES

1. M. J. Schick (Ed.), *Nonionic Surfacants*, Surfactant Science Series, Vol. 1, Marcel Dekker, New York, 1966.

2. J. Cross (Ed.), *Nonionic Surfacants: Chemical Analysis*, Surfactant Science Series, Vol. 19, Marcel Dekker, New York, 1987.
3. M. J. Schick (Ed.), *Nonionic Surfacants: Physical Chemistry*, Surfactant Science Series, Vol. 23, Marcel Dekker, New York, 1987.
4. T. M. Schmitt, (Ed.), *Analysis of Surfactants*, Surfactant Science Series, Vol. 40, Marcel Dekker, New York, 1992.
5. American Society for Testing and Materials, Standard Test Method for Cloud Point of Nonionic Surfactants, D2024–65. 1916 Race St., Philadelphia, PA.
6. R. M. Ianniello, *Anal. Lett. 21*:87 (1988).
7. American Society for Testing and Materials, Standard Test Method for Ash from Petroleum Products, D482–87. 1916 Race St., Philadelphia, PA.
8. J. G. W. Chlebicki, *Tenside Surf. Deterg. 15*:187 (1978).
9. J. G. W. Chlebicki, *Tenside Surf. Deterg. 17*:13 (1980).
10. M. J. Rosen, *Anal. Chem. 35*:2074 (1963).
11. J. J. Hejna and D. Daly, *J. Soc. Cosmetic Chem. 21*:107 (1970).
12. M. Zgoda and S. Petri, *Chem. Anal. (Warsaw) 31*:577 (1986).
13. M. E. Leon-Gonzalez, M. J. Santos-Delgado, and L. M. Polo-Diez, *Analyst 115*:609 (1990).
14. G. Fleischer, P. Bloss, and W-D. Hergeth, *Coll. Polym. Sci. 271*:217 (1993).
15. J. Rassing, W. P. McKenna, S. Bandyopadhyay, and E. M. Eyring, *J. Mol. Liq. 27*:165 (1984).
16. K. Nakamura, R. Endo, and M. Takeda, *J. Polym. Sci. Polym. Phys. Ed. 15*:2095 (1977).
17. J. Juhasz, V. Lenaerts, P. V. Minh Tan, and H. Ong, *J. Colloid Interface Sci. 136*:168 (1990).
18. P. Bahadur, K. Pandya, M. Almgren, P. Li, and P. Stilbs, *Coll. Polym. Sci. 271*:657 (1993).
19. Z. Zukang and B. Chu, *Macromolecules 20*:3089 (1987).
20. G. Wanka, H. Hoffman, and W. Ulbricht, *Coll. Polym Sci. 268*:101 (1990).
21. K. Schillen, W. Brown, and C. Konak, *Macromolecules 26*:3611 (1993).
22. K. Schillen, O. Glatter, and W. Brown, *Prog. Coll. Polym. Sci. 93*:66 (1993).
23. O. Glatter, K. Schillen, and G. Scherf, *Prog. Coll. Polym. Sci. 93*:251 (1993).
24. L. Yang, A. D. Bedelis, D. Attwood, and C. Booth, *J. Chem. Soc., Faraday Trans. 88*:1447 (1992).
25. M. M. Amiji and K. Park, *J. Appl. Polym. Sci. 52*:539 (1994).
26. N. Funasaki, S. Hada, and S. Neya, *J. Phys. Chem. 92*:7112 (1988).
27. W-B. Sun, J-F. Ding, R. H. Mobbs, D. Attwood, and C. Booth, *Colloids Surfaces 54*:103 (1991).
28. Q. Wang, C. Price, and C. Booth, *J. Chem. Soc., Faraday Trans. 88*:1437 (1992).

29. Q. Wang, G-E. Yu, Y. Deng, C. Price, and C. Booth, *Eur. Polym. J.* *29*:665 (1993).
30. I. Alig, R. U. Ebert, W. D. Hergeth, and S. Wartewig, *Polym. Comm.* *31*:314 (1990).
31. N. M. Mitchard, A. E. Beezer, J. C. Mitchell, J. K. Armstrong, B. Z. Chowdhry, S. Leharne, and G. Buckton, *J. Phys. Chem.* *96*:9507 (1992).
32. Y. Deng, G-E. Yu, C. Price, and C. Booth, *J. Chem. Soc., Faraday Trans.* *88*:1441 (1992).
33. M. W. Edens, V. M. Nace, and J. C. Knoell, *Polymeric Materials Science and Engineering*, Proceedings of the ACS Division of Polymeric Materials Science and Engineering, Vol. 69, American Chemical Society, Washington, D.C., 1993, p. 551.
34. H. Koenig, *Fresenius' Z. Anal. Chem.* *251*:225 (1970).
35. H. Koenig, *Tenside Surf. Deterg.* *8*:63 (1971).
36. G. Carminati, L. Cavalli, and F. Buosi, *J. Am. Oil Chem. Soc.* *65*:669 (1988).
37. A. J. Montana, in *Nonionic Surfactants: Chemical Analysis* (J. Cross, Ed.), Surfactant Science Series, Vol. 19, Marcel Dekker, New York, 1987.
38. W. Gronski, G. Hellmann, and A. Wilsch-Irrgang, *Makromol. Chem.* *192*:591 (1991).
39. C. L. LeBas and P. A. Turley, *J. Cell Plast.* *20*:194 (1984).
40. M. M. Zgoda, *Acta Pol. Pharm.* *45*:63 (1988).
41. D. M. Gabriel, *J. Soc. Cosmet. Chem.* *25*:33 (1974).
42. C. G. Naylor, *J. Am. Oil Chem. Soc.* *63*:1201 (1986).
43. American Society for Testing and Materials, Standard Methods for Testing Polyurethane Raw Materials, D4273–83. 1916 Race St., Philadelphia, PA.
44. Society of the Plastics Industry, Standard Method for Testing Polyurethane Raw Materials: Determination of the Polymerized Ethylene Oxide Content of Polyether Polyols, PURMAC-5.
45. F. Heatley, Y-Z. Luo, J-F. Ding, R. H. Mobbs, and C. Booth, *Macromolecules* *21*:2713 (1988).
46. E. B. Whipple and P. J. Green, *Macromolecules* *6*:38 (1973).
47. X-C. Chen, *Bopoxue Zazhi* *1*:409 (1984).
48. F. Halmo, L. Malik, and T. Liptaj, *Chem. Prum.* *36*:203 (1986).
49. D. L. Rubenstein and T. T. Nakashima, *Anal. Chem.* *51*:1465A (1979).
50. C. Lecocq and J-Y. Lallemand, *J. Chem. Soc., Chem. Commun.* *4*:150 (1981).
51. A. G. Ferrige and J. C. Lindon, *J. Magn. Reson.* *31*:337 (1978).
52. S. L. Patt and J. N. Shoolery, *J. Magn. Reson.* *46*:535 (1982).
53. J. L. Carrion and M. De la Guardia, *Quim. Anal. (Barcelona)* *6*:76 (1987).
54. G. Rauscher, in *Nonionic Surfactants: Chemical Analysis* (J. Cross, Ed.), Surfactant Science Series, Vol. 19, Marcel Dekker, New York, 1987.

55. M. P. Fuller, G. L. Ritter, and C. S. Draper, *App. Spectrosc. 88*:228 (1988).
56. M. P. Fuller and M. E. Meyers, *Mikrochem. Acta 1*:31 (1988).
57. N. A. Puttman, S. Lee, and B. H. Baxter, *J. Soc. Cosm. Chem. 16*:607 (1965).
58. P. A. Turley and A. Pietroantonio, *J. Cell. Plast. 20*:274 (1984).
59. C. V. London and K. S. Krishnan, *Nature 121*:501 (1928).
60. P. R. Carey, *Biochemical Applications of Raman and Resonance Raman Spectroscopies*, Academic Press, Boston, 1982.
61. R. J. H. Clark and R. E. Hester, *Advances in Infrared and Raman Spectroscopy*, Hayden and Son, London, 1980.
62. D. A. Long, *Chem. Brit. 25*:589 (1989).
63. W. H. Washburn, *Amer. Lab. Nov.*:48 (1978).
64. S. M. Mason, N. Conroy, N. M. Dixon, and K. P. J. Williams, *Spectrochim. Acta 49A*:633 (1993).
65. T. Hirschfeld and B. Chase, *Appl. Spectrosc. 40*:133 (1986).
66. P. Hendra, C. Jones, and G. Warnes, *Fourier Transform Raman Spectroscopy*, Ellis Horwood Ltd., Chichester, UK, 1991.
67. J. L. Koenig, *Appl. Spectrosc. Rev. 4*:233 (1971).
68. J. J. Gagel and K. Biemann, *Anal. Chem. 59*:1266 (1987).
69. R. P. Markovich, L. G. Hazlitt, and L. Smith-Courtney, in *Chromatography of Polymers: Characterization by SEC and FFF* (T. Provder, Ed.), ACS Symposium Series 521, American Chemical Society, Washington, D.C., 1993.
70. L. M. Wheeler and J. N. Willis, *Appl. Spectrosc. 47*:1128 (1993).
71. A. H. Dekmezian and T. Morioka, *Anal. Chem. 61*:458 (1989).
72. G. Weiss, *Fette, Seifen, Anstrichm. 70*:355 (1968).
73. N. V. Chukanov, I. V. Kumpanenko, K. S. Kazanskii, and S. G. Entelis, *Polym. Sci. USSR 18*:3105 (1976).
74. R. Janik and J. Plucinski, *Chem. Anal. (Warsaw) 26*:29 (1981).
75. E. A. Burns and R. F. Muraca, *Anal. Chem. 31*:397 (1959).
76. C. L. Hilton, *Anal. Chem. 31*:1610 (1959).
77. C. S. Y. Kim, A. L. Dodge, S. Lau, and A. Kawasaki, *Anal. Chem.* 54: 232 (1982).
78. R. E. Schirmer and A. G. Gargus, *Am. Lab. Nov.*:37 (1988).
79. H. T. Kalinoski, in *Cationic Surfactants: Analytical and Biological Evaluation* (J. Cross and E. J. Singer, Eds.), Surfactant Science Series, Vol. 53, Marcel Dekker, New York, 1994.
80. A. K. Lee and R. D. Sedgwick, *J. Polym. Sci., Polym. Chem. Ed. 16*:685 (1978).
81. A. Zeman, *Angew. Makromol. Chem. 31*:1 (1973).

82. S. A. Liebman and E. J. Levy, Eds., *Pyrolysis and GC in Polymer Analysis*: Chromatographic Science Series, Vol. 29, Marcel Dekker, New York, 1985.
83. J. C. Tou, D. Zakett, and V. J. Caldecourt, in *Tandem Mass Spectrometry* (F. W. McLafferty, Ed.), J. Wiley and Sons, New York, 1983.
84. R. P. Lattimer, H. Münster, and H. Budzikiewicz, *Int. J. Mass Spectrom. Ion Processes 90*:119 (1989).
85. H. T. Kalinoski and L. O. Hargiss, *J. Am. Soc. Mass Spectrom. 3*:150 (1992).
86. R. P. Lattimer, *J. Am. Soc. Mass Spectrom. 3*:225 (1992).
87. R. P. Lattimer, *J. Am. Soc. Mass Spectrom. 5*:1072 (1994).
88. T. S. Selby, C. Wesdemiotis, and R. P. Lattimer, *J. Am. Soc. Mass Spectrom. 5*:1081 (1994).
89. M. Yamashita and J. B. Fenn, *J. Phys. Chem. 88*:4451 (1984).
90. R. D. Smith, J. A. Olivares, N. T. Nguyen, and H. R. Udseth, *Anal. Chem. 60*:436 (1988).
91. M. Karas and F. Hillenkamp, *Anal. Chem. 60*:2299 (1988).
92. J. B. Fenn, M. Mann, C. K. Meng, S. F. Wong, and C. M. Whitehouse, *Mass Spectrom. Rev. 9*:37 (1990).
93. E. C. Huang, T. Wachs, J. J. Conboy, and J. D. Henion, *Anal. Chem. 62*:713A (1990).
94. M. Karas, U. Bahr, and U. Giessmann, *Mass Spectrom. Rev. 10*:335 (1991).
95. M. Karas and U. Bahr, *Trends Anal. Chem. 9*:321 (1990).
96. F. Hillenkamp, M. Karas, R. C. Beavis, and B. T. Chait, *Anal. Chem. 63*:1193 (1991).
97. C. N. McEwen, B. S. Larsen, and W. J. Simonsick, Jr., *Proceedings of the 42nd ASMS Conference on Mass Spectrometry and Allied Topics*, Chicago, 1994, p. 317.
98. L. Prokai and W. J. Simonsick, Jr., *Rapid Commun. in Mass Spectrom. 7*:853 (1993).
99. W. J. Simonsick, Jr. and L. Prokai, *Proceedings of the 42nd ASMS Conference on Mass Spectrometry and Allied Topics*, Chicago, 1994, p. 318.
100. F. W. McLafferty, *Acc. Chem. Res. 27*:379 (1994).
101. P. B. O'Conner, J. P. Speir, M. W. Senko, D. P. Little, and F. W. McLafferty, *J. Mass Spectrom. 30*:88–93 (1995).
102. W. C. Wiley and I. H. McLaren, *Rev. Sci. Instrum. 26*:1150 (1955).
103. R. J. Cotter, *Anal. Chem. 64*:1027A (1992).
104. J. Claereboudt, M. Claeys, H. Geise, R. Gijbels, and A. Vertes, *J. Am. Soc. Mass Spectrom. 4*:798 (1993).
105. U. Bahr, M. Karas, F. Hillenkamp, and U. Giessmann, *Anal. Chem. 64*:2866 (1992).

106. M. Karas, A. Deppe, F. Hillenkamp, and U. Giessmann, *Proceedings of the 42nd ASMS Conference on Mass Spectrometry and Allied Topics*, Washington, D.C., 1992, p. 291.
107. J. A. Castro, C. Koester, and C. Wilkins, *Rapid Commun. Mass Spectrom. 6*:239 (1992).
108. L-S. Sheng, S. Shew, J. Covey, B. Winger, and J. Campana, *Proceedings of the 42nd ASMS Conference on Mass Spectrometry and Allied Topics*, Chicago, 1994, p. 320.
109. G. B. Kenion, R. A. Ludicky, J. C. Via, and D. E. Doster, *Proceedings of the 42nd ASMS Conference on Mass Spectrometry and Allied Topics*, Chicago, 1994, p. 319.
110. International Organization for Standardization, ISO 2270, Geneva, Switzerland (1989).
111. S. Siggia, A. C. Starke, Jr., J. Garis, Jr., and C. R. Stahl, *Anal. Chem. 30*:115 (1958).
112. T. Provder, Ed., *Chromatography of Polymers: Characterization by SEC and FFF*, ACS Symposium Series 521, American Chemical Society, Washington, D.C., 1993.
113. Y. Deng, C. Price, and C. Booth, *Eur. Polym. J. 30*:103 (1994).
114. H. Benoit, P. Rempp, and Z. Grubisic, *J. Polym. Sci., Phys. Ed. 5*:753 (1967).
115. W. Bahary and M. Jilani, *J. Appl. Polym. Sci. 48*:1531 (1993).
116. W. W. Yau, J. Kirkland, and D. D. Bly, *Modern Size Exclusion Liquid Chromatography*, J. Wiley and Sons, New York, 1979.
117. J. V. Dawkins, N. P. Gabbott, A. Montenegro, L. Loyd, and F. Warner, *J. Appl. Polym. Sci. Polym. Symp. 45*:103 (1990).
118. R. A. Sanayei, K. F. O'Driscoll, and A. Rudin, in *Chromatography of Polymers: Characterization by SEC and FFF* (T. Provder, Ed.), ACS Symposium Series 521, American Chemical Society, Washington, D.C., 1993.
119. F. Gores and P. Kilz, in *Chromatography of Polymers: Characterization by SEC and FFF* (T. Provder, Ed.), ACS Symposium Series 521, American Chemical Society, Washington, D.C., 1993.
120. T. H. Mourey and S. T. Balke, in *Chromatography of Polymers: Characterization by SEC and FFF* (T. Provder, Ed.), ACS Symposium Series 521, American Chemical Society, Washington, D.C., 1993.
121. B. G. Belenkii and E. S. Gankina, *J. Chromatogr. 141*:13 (1977).
122. P. P. Nefedov and T. P. Zhmakina, *Vysokomol. soedin. A23*:276 (1981).
123. D. Hunkeler, T. Macko, and D. Berek, in *Chromatography of Polymers: Characterization by SEC and FFF* (T. Provder, Ed.), ACS Symposium Series 521, American Chemical Society, Washington, D.C., 1993.
124. T. M. Zimina, E. E. Kever, E. Y. Melenevskaya, V. N. Egonnik, and B. G. Belenkii, *Vysokomol. soedin. A33*:1349 (1991).

125. A. V. Gorshkov, V. V. Evreinov, and S. G. Entelis, *Zh. fiz. khim. 62*:490 (1988).
126. N. N. Filatova and A. V. Gorshkov, *Vysokomol. soedin. A30*:953 (1980).
127. M. N. Myers, P. Chen, and J. C. Giddings, in *Chromatography of Polymers: Characterization by SEC and FFF* (T. Provder, Ed.), ACS Symposium Series 521, American Chemical Society, Washington, D.C., 1993.
128. M. E. Schimpf, L. M. Wheeler, and P. F. Romeo, in *Chromatography of Polymers: Characterization by SEC and FFF* (T. Provder, Ed.), ACS Symposium Series 521, American Chemical Society, Washington, D.C., 1993.
129. P. Voogt, *Fette, Seifen, Anstrichm. 65*:964 (1963).
130. A. Mathias and N. Mellor, *Anal. Chem. 38*:472 (1966).
131. J. B. Stead and A. H. Hindley, *J. Chromatogr. 42*:470 (1969).
132. P. W. Morgan, *Ind. Eng. Chem., Anal. Ed. 18*:500 (1946).
133. F. W. Cheng, *Microchem. J. 9*:270 (1965).
134. K. Obruba, E. Kucerova, and M. Juracek, *Mikrochim. Ichnoanal. Acta*, 44 (1964).
135. K. L. Hodges, W. E. Kester, D. L. Wiederrich, and J. A. Grover, *Anal. Chem. 51*:2172 (1979).
136. American Society for Testing and Materials, Standard Test Method for Ethylene Oxide Content of Polyethoxylated Nonionic Surfactants, D2959–76. 1916 Race St., Philadelphia, PA.
137. J. Szymanowski, P. Kusz, and E. Dziwinski, *J. Chromatogr. 455*:131 (1988).
138. P. Waszeciak and H. G. Nadeau, *Anal. Chem. 36*:764 (1964).
139. P. Kusz, J. Szymanowski, K. Pyzalski, and E. Dziwinski, *LC-GC 8*:48 (1990).
140. K. Tsuji and K. Konishi, *J. Am. Oil Chem. Soc. 52*:106 (1975).
141. K. Tsuji and K. Konishi, *J. Am. Oil Chem. Soc. 51*:55 (1974).
142. K. Tsuji and K. Konishi, *Analyst 96*:457 (1971).
143. I. Zeman, L. Novak, L. Mitter, J. Stekla, and O. Holendova, *J. Chromatogr. 119*:581 (1976).
144. S. T. Hirozawa, in *Treatise on Analytical Chemisty, Part II* (I. M. Kolthoff and P. J. Elving, Eds.), J. Wiley and Sons, New York, 1971.
145. American Society for Testing and Materials, Standard Test Method for Total Bromine Number of Unsaturated Aliphatic Chemicals, E234–78. 1916 Race St., Philadelphia, PA.
146. American Society for Testing and Materials, Standard Test Method for Vinyl Unsaturation in Organic Compounds with Mercuric Acetate, E441–74. 1916 Race St., Philadelphia, PA.
147. American Society for Testing and Materials, Standard Test Method for Iodine Value of Drying Oils and Fatty Acids, D1959–85. 1916 Race St., Philadelphia, PA.

148. International Agency for Research on Cancer, *Monograph on the Evaluation of Carcinogenic Risks of Chemicals to Humans, Suppl. 7*: 201 (1987).
149. National Cancer Institute, Carcinogenicity Technical Report No. NCI-C G-TR-80, Bethesda, MD (1978).
150. L. Fishbein, in *Potential Industrial Carcinogens and Mutagens*, Studies in Environmental Science, Vol. 4., Elsevier, New York 1979.
151. I. Lundberg, J. Hogberg, T. Kronevi, and B. Holmberg, *Cancer Lett. 36*:29 (1987).
152. G. D. Clayton and F. E. Clayton, Eds., *Patty's Industrial Hygiene and Toxicology, Vol. 2C*, J. Wiley and Sons, New York 1982.
153. A. Fairly, E. C. Clinton, A. H. Ford-Moore, *J. Hyg. (London) 34*:486 (1934).
154. *United States Pharmacopeia, 22nd Revision/National Formulary*, 17th Ed., U.S. Pharmacopeial Convention, Inc., Rockville, MD, 1989.
155. T. J. Birkel, C. R. Warner, and T. Fazio, *J. Assoc. Off. Anal. Chem. 62*:931 (1979).
156. S. C. Rastogi, *Chromatographia 29*:441 (1990).
157. J. Rümenapp and J. Hild, *Lebensmittelchem. Gerichte. Chem. 41*:59 (1987).
158. J. R. Dahlgran and C. R. Shingleton, *J. Assoc. Off. Anal. Chem. 70*:796 (1987).
159. S. Scalia, M. Guarneri, and E. Menegatti, *Analyst 115*:929 (1990).
160. D. B. Black, R. C. Lawrence, E. G. Lovering, and J. R. Watson, *J. Assoc. Off. Anal. Chem. 66*:180 (1983).
161. J. R. Dahlgran and M. N. Jameson, *J. Assoc. Off. Anal. Chem. 71*:560 (1988).
162. R. D. Mair and R. T. Hall, in *Treatise on Analytical Chemisty, Part II* (I. M. Kolthoff and P. J. Elving, Eds.), J. Wiley and Sons, New York, (1971).
163. R. M. Johnson and I. W. Siddiqi, *Determination of Organic Peroxides*, Pergamon Press, New York, 1970.
164. American Society for Testing and Materials, Standard Test Methods for Assay of Organic Peroxides E298–84. 1916 Race St., Philadelphia, PA.
165. American Society for Testing and Materials, Standard Test Method for Trace Amounts of Peroxides in Organic Solvents, E299–84. 1916 Race St., Philadelphia, PA.
166. E. Azaz, M. Donbrow, and R. Hamburger, *Analyst 98*:663 (1973).
167. American Society for Testing and Materials, Standard Test Method for Trace Quantities of Carbonyl Compounds with 2,4-Dinitrophenylhydrazine, E411–70. 1916 Race St., Philadelphia, PA.
168. G. R. Lappin and L. C. Clark, *Anal. Chem. 23*:541 (1951).
169. American Society for Testing and Materials, Standard Test Method for Determination of Carbonyls in C4 Hydrocarbons, D4423–84. 1916 Race St., Philadelphia, PA.

170. T. L. Chester, J. D. Pinkston, and D. E. Raynie, *Anal. Chem.* *66*:106R (1994).
171. S. A. Westwood, Ed., *Supercritical Fluid Extraction and its Use in Chromatographic Sample Preparation*, CRC Press, Boca Raton, FL, 1993.
172. J. H. Braybrook and G. A. MacKay, *Polym. Int.* *27*:157 (1992).
173. B. E. Richter, D. E. Knowles, J. L. Ezzell, and N. L. Porter, *Natl. Meet. Am. Chem. Soc., Div. Environ. Chem.* *33*:347 (1993).
174. F. Hoefler, *Labor Praxis* *16*:506 (1992).
175. J. M. Levy, A. C. Rosselli, E. Storozynsky, R. Ravey, L. A. Dolata, and M. Ashraf-Khorassani, *LC-GC* *10*:386 (1992).
176. S. B. Hawthorne, in *Analysis with Supercritical Fluids: Extraction and Chromatography* (B. Wenclawiak, Ed.), Springer-Verlag, Berlin, 1992.
177. M. T. G. Hierro and G. Santa-Maria, *Food Chem.* *45*:189 (1992).
178. M. H. Liu, S. Kapila, K. S. Nam, and A. A. Elseewi, *J. Chromatogr.* *639*:151 (1993).
179. J. M. Levy and M. Ashraf-Khorassani, *J. Chromatogr. Libr.* *53*:197 (1992).
180. G. A. MacKay and R. M. Smith, *Anal. Proc.* 29: 463 (1992).
181. L. Baner, T. Buecherl, J. Ewender, and R. Franz, *J. Supercrit. Fluids* *5*:213 (1992).
182. H. Daimon and Y. Hirata, *Chromatographia* *32*:549 (1991).
183. M. Ashraf-Khorassani, D. S. Boyer, and J. M. Levy, *J. Chromatogr. Sci.* *29*:517 (1991).
184. T. P. Hunt, C. J. Dowle, and G. Greenway, *Analyst (Lond.)* *116*:1299 (1991).
185. T. P. Hunt, C. J. Dowle, and G. Greenway, *Analyst (Cambridge, UK)* *118*:17 (1993).
186. H. T. Kalinoski and A. Jensen, *J. Am. Oil Chem. Soc.* *66*:1171 (1989).
187. B. Weibull, *Vortaege Originalfassung Intern. Kongr. Grenzflaechenaktive Stoffe, 3*, Cologne, 1960, p. 121.
188. K. Buerger, *Fresenius' Z. Anal. Chem.* *196*:22 (1963).
189. M. Coupkova, K. Janes, J. Sanitrak, and J. Coupek, *J. Chromatogr.* *160*:73 (1978).
190. H. Szewczyk and J. Szmanowski, *Tenside Surf. Deterg.* *19*:357 (1982).
191. I. Zeman and M. Paulovic, in *Proceedings of the 2nd World Surfactants Congress*, Paris, 1988, p. 384.
192. W. Winkle, *Chromatographia* *29*:530 (1990).
193. C. A. Eckert, M. P. Ekart, B. L. Knutson, K. P. Payne, D. L. Tomasko, C. L. Liotta, and N. R. Foster, *Ind. Eng. Chem. Res.* *31*:1105 (1992).
194. M. Daneshvar and E. Gulari, *J. Supercrit. Fluids* *5*:143 (1992).

3

Physical Chemistry of Polyoxyalkylene Block Copolymer Surfactants

BENJAMIN CHU and ZUKANG ZHOU Department of Chemistry, State University of New York at Stony Brook, Stony Brook, New York

I.	Introduction	68
II.	Recent Studies of Pluronic Polyols	69
III.	Physical Methods	69
IV.	Association Characteristics of EO/PO Polymeric Surfactants in the Dilute Regime	75
	A. Critical micelle concentration and critical micelle temperature	75
	B. Enthalpy of micellization and entropy-driven nature of the selfassembly process	82
	C. Association number, micelle size, and micellar structure	89
	D. Micellar shape transition	93
	E. Composition inhomogeneity and anomalous micellization	95
	F. Effect of chain architecture	98
	G. Surface activity and surface adsorption	102
	H. Solubilization behavior	110
V.	Phase Behavior of EO/PO Polymeric Surfactants in Water	112
	A. Introductory remarks	112
	B. Phase diagrams	112
	C. Small-angle neutron scattering studies on phase behavior and phase structure	118

D. Chain architecture and phase behavior 122
E. Phase behavior in nonaqueous solvents 126

VI. Gelling Properties of EO/PO Polymeric Surfactants in Water 126
A. Thermoreversible gel formation 126
B. Phase diagrams in terms of unimer, micelle, and gel regions 128
C. Effect of additives on gel region 131

VII. Micelle Formation and Gelation of EO/BO Polymeric Surfactants 133

VIII. Theoretical Development 135

References 136

I. INTRODUCTION

Water-soluble triblock copolymers of the EPE-type [P and E represent poly(oxypropylene) and poly(oxyethylene), respectively], often abbreviated as POE-POP-POE [or $(EO)_a(PO)_b(EO)_c$ with a, b, c representing the number of repeat units] are commercially available non-ionic macromolecular surfactants and have many uses. In a recent review article on poloxamers, Schmolka [1,2] stated that there were over 1000 articles on many applications in the medical and pharmaceutical industries alone. The commercial names for these triblocks, which find widespread industrial applications in their uses as emulsifying, wetting, thickening, coating, solubilizing, stabilizing, dispersing, lubricating, and foaming agents, are Pluronic® (BASF) or Synperonic® (ICI) polyols. A variety of the EPE-type triblocks are available with varying chain length (total copolymer molecular weight), block length ratio, and chain architecture (e.g., PEP-type, instead of EPE-type). In addition to the EPE-type triblocks, EB-type diblocks and EBE-type triblocks—with B denoting poly(oxybutylene) and being more hydrophobic than E—have been synthesized and studied, and are being marketed by The Dow Chemical Co. The Pluronic polyols are made up of relatively shorter segment blocks. Nevertheless, these nonionic macromolecular surfactants have many properties similar to those of long-chain triblock copolymers. Micelles of block and graft copolymers in selective solvents have been reviewed extensively by Tuzar and Kratochvil [3]. The physical chemistry of nonionic surfactants on earlier studies have also been presented [4]. The main physical methods used in studying Pluronic polyols have been reviewed by Chu [5].

The aims of this article are to focus on the physical chemistry of block copolymer association by the EO, PO, and BO blocks, including a broad background of references. The chapter cannot possibly cover the results of all the authors, but tries to summarize, from a subjective viewpoint, some useful conclusions that are common to many of the literature citations. More specifically, we provide an outline of basic studies which have been reported by different research groups. We hope that the reader can quickly gather a sense as to who are the players in the game. The physical methods used to measure the association behavior and the micellar structure are outlined. Finally, selective properties, such as the critical micelle concentration (cmc) and association number, have been reformulated in tabular form for ease of reference. The approach to this review is different from the earlier paper [5] where emphasis was based on the work done at Stony Brook. Furthermore, we shall emphasize the review mainly on two-component systems. The physical properties of Pluronic polyols in water/xylene mixtures are outlined briefly towards the end of this chapter.

II. RECENT STUDIES OF PLURONIC POLYOLS

There are many research groups studying the basic properties of Pluronic polyols. Some constitute fairly large teams, while others represent only a few individuals. We arbitrarily group the activities together based on location, which could include more than one independent group. Table 1 lists the activities which have been reported in recent years, including the main physical methods used. From Table 1, one notes that there are more than ten research groups involved in studying the Pluronic polyols. Many different physical techniques have been used to study a large variety of the Pluronic polyols. Unfortunately, most research groups are limited in the range of techniques available at their disposal. Thus, the viewpoints are often limited. The comparison of data has been made difficult because different polyols were used for measurements under different experimental conditions. Some theory [123,124,167] and simulations [125,126] are beginning to emerge. However, it is a challenge to make comparisons with experiments, often for lack of information on the precise values of physical parameters needed to test the theory. Furthermore, simulations have been limited to block copolymers with few repeat units. The theoretical development by Linse deals directly with the polyols and has obtained good agreement with experiments.

III. PHYSICAL METHODS

There are many physical methods used to study the Pluronic polyols. Some specialized techniques have marginal value and provide only a limited amount of narrowly focused information. We summarize the main physical techniques under the headings of scattering, spectroscopic, and other techniques in Table 2.

TABLE 1 Research Activities on Pluronic Polyols (and Some Other Triblocks)

Location	Main physical methods used	Refs.
1. University of Kent at Canterbury, U.K.	HSDSC	6–11
2. Bayreuth, Germany	SLS, DLS, DSC, TEB, SANS, NMR, phase behavior, surface tension	12–16
3. Uppsala, Sweden	SLS, DLS, rheology, ultrasonic velocimetry, fluorescence, Brillouin scattering, NMR, SAXS	17–29
4. South Gujarat University, Surat, India	Cloud point, sedimentation, viscosity, LS, sound velocity, PGSE-NMR	30–33
5. Riso, Denmark	Phase diagram, SANS, SLS, DLS, rheology	34–43
6. Lund, Sweden	Adsorption, phase behavior, PGSE-NMR, theory, NMR, light transmittance	44–62
7. MIT, Cambridge, MA	Dye solubilization, surface tension	63–67
8. Manchester, U.K.	Surface tension, SLS, DLS, DSC, GPC, NMR, EGPC, ultrasonic velocity	68–93
9. Technische Universitat Merseburg, Germany	Ultrasonic absorption, IR and Raman spectroscopy, viscosity, LS, absorption	94–97
10. University of Leipzig, Germany	Viscosity, PGSE-NMR	98–100
11. SUNY Stony Brook, NY	Phase behavior, SLS, DLS, SAXS, SANS, TEB, viscosity, NMR	101–117
12. Columbia University, NY	Photoluminescent probes	118–121
13. Pennsylvania State Univ., University Park, PA	Theory, GC	122–124
14. Akron, OH	Simulation	125,126
15. Individuals		127–143

It is too lengthy to provide a summary of equations which represent the experimental quantities measured and the information which can be derived thereof. Instead, the key experimental quantities measured and the information derived are listed in Table 2. Selective references have been provided for the reader who is interested in taking a closer look at the techniques.

The scattering techniques involve the use of a probe radiation which can be (visible) laser light, X-rays, or neutrons, as we are dealing with sizes in the nanometer scale to those approaching the wavelength of light. The investigation of size, shape, and internal structure comes from measurements of the interactions between the particles and the probing radiation. In order to be able to vary the magnitude of the scattering wave vector q over a large range, one chooses light, X-rays or neutrons depending on the size range of interest. Here,

$$q = \left(\frac{4\pi}{\lambda}\right)\sin\left(\frac{\theta}{2}\right)$$

with λ and θ being the wavelength of probing radiation and the scattering angle, respectively. For a comparison of light, X-rays, and neutrons, the reader is referred to [144]. Briefly, light has the longest wavelength and hence is limited to the low q range, implying that we can use it to measure the weight-average molecular weight (M_w) from extrapolated zero-angle time-averaged absolute scattered intensity at infinite dilution, the second virial coefficient (A_2) from extrapolated zero-angle time-averaged scattered intensity measured as a function of concentration, and the radius of gyration (R_g) from the angular distribution of time-averaged scattered intensity extrapolated to infinite dilution. It should be noted that we are dealing with block copolymers. Thus, we cannot obtain a true molecular weight and a true radius of gyration because the refractive index increments of the blocks are likely to be different. Fortunately, for the EO, PO, and BO blocks, the differences are very small and the corrections amount to only a few percent. Nevertheless, we should accept the results as quantitative within this approximation. At higher concentrations, the block copolymers can form micelles and infinite dilution is now referred to the critical micelle concentration. By knowing the initial molecular weight of the polymer, we can then compute the association number (n_w) from the measured molar mass of the micelle. Again, such computations have neglected the polydispersity effect. Small angle X-ray scattering (SAXS) and small angle neutron scattering (SANS) offer similar types of information as static light scattering (SLS), except that the refractive index increment in SLS is replaced by the electron density difference and the scattering length difference, respectively. The reader may be interested to know that the sensitivity is greatly reduced for SAXS and SANS since these probing beams have a lower power density. Consequently, many SAXS and SANS experiments cannot be performed in the dilute regime. In

TABLE 2 Physical Methods for Characterization of Block Copolymers in Solution

Methods	Experimental quantities measured	Information derived
Scattering techniques		
Static light scattering (SLS) [144]	Time-averaged scattered intensity	Weight-average molecular weight (M_w), second virial coefficient (A_2), and radius of gyration (R_g); association number of micelle (n_w); critical micelle temperature (cmt); critical micelle concentration (cmc); equivalent hard sphere radius R_{hs}, if appropriate, either from M and A_2 or by the model fitting of the I versus C curve with I and C being the scattered intensity and the polymer concentration, respectively; core radius (r_c) from n_w.
Dynamic light scattering (DLS) [144]	Fluctuation of scattered intensity with time, i.e. intensity-intensity autocorrelation function	Translational diffusion coefficient D_t (z-average) and in some cases rotational diffusion coefficient D_r; hydrodynamic radius (R_h); particle size distribution in terms of R_h; estimation of weight fractions of unimer (u) and micelle (m) species with known n_w and I_m/I_u; internal structure from R_g/R_h; shell thickness (L_s) from R_h and r_c.
Small angle X-ray scattering (SAXS) [145,146]	Similar as SLS, but covers higher q values	R_g and internal structure. unimer/micelle size and shape by the fit to simple geometrical models; detection and structural parameters of ordered mesophases at high concentrations
Small angle neutron scattering (SANS) [162]	Similar as SAXS, but with controlled contrast variation	Similar as SAXS via H-D substitution

TABLE 2 (Continued)

Methods	Experimental quantities measured	Information derived
Spectroscopic techniques		
NMR spectroscopy	Chemical shift	Local environment
Pulsed-gradient spin-echo nuclear magnetic resonance (PGSE-NMR)	Echo amplitude decay as a function of field gradient duration	Selfdiffusion coefficient D (weight average) and R_h similar as DLS.
Fluorescence probing	Fluorescence spectra, polarization and/or fluorescence decay	Association number; micropolarity; microviscosity.
Dye solubilization	Light absorbance	Critical micelle concentration.
Other techniques		
Viscosimetry	Solution viscosity	Intrinsic viscosity [η]; hydrodynamic volume of particle, e.g. R_η, from M and [η]; shape evaluation from concentration dependence of viscosity.
Surface tension	Surface tension of solution against air	Critical micelle concentration; surface activity and surface adsorption; area occupied per copolymer chain at the interface. composition impurity detection by the existence of a minimum.
Ultrasound velocimetry	Speed of sound	Critical micelle concentration; critical micelle temperature.
Differential scanning calorimetry (DSC)	Endothermic or exothermic heat	Critical micelle temperature region; enthalpy of micellization; phase transition.
Eluent gel permeation chromatograpy (EGPC)	Eluent volume when using block copolymer as eluent solution	Association equilibrium of unimers and micelles; critical micelle temperature.

order to correct for the intermolecular interactions, a structure factor based on some model has to be used. Thus, the resultant form factor contains the approximations involved in the structure factor. A careful examination of the effects in different q ranges is advisable if one aims to probe deeper into the internal structures of the micelle.

In dynamic light scattering (DLS), the intensity-intensity time correlation function $G^{(2)}(t)$ can yield information on the translational diffusion coefficient (D_t) and, in some cases, the rotational diffusion coefficient (D_r for anisotropic particles). From D_t, and with the aid of the Stokes-Einstein relation, we can compute the hydrodynamic radius (R_h). Laplace inversion of $G^{(2)}(t)$ yields information on the characteristic linewidth (and the resulting hydrodynamic size) distribution. In a mixture of unimers (u) and micelles (m), the intensity contributions (I_u and I_m) coming from the bimodal size distribution, together with the association number, can provide us an estimate of weight fractions of unimer and micelle species. Some information on the compactness of the spherical micelles formed can be estimated by considering the ratio of R_g/R_h.

Proton and ^{13}C nuclear magnetic resonance (NMR) spectroscopy permit us to examine the local environment of EO, PO, and BO chains. In some cases, we can detect how the water is bound to the EO segment or whether the conformation of the carbon chains changes upon micellization. Pulsed-gradient spin-echo (PGSE) NMR can be used to measure the selfdiffusion coefficient (and R_h) of the micelles. It is less common, but its utility should not be ignored.

Fluorescence probes and dye solubilization are sensitive techniques to determine the association number and cmc. The techniques, on the other hand, involve knowledge of photophysics and photochemistry related to interactions of the dye (or fluorescence probes) with block copolymers as well as selective solvents.

The more classical techniques to study micellar systems deal with measurements of macroscopic properties, such as viscosity and surface tension, or thermodynamic properties as determined by differential scanning calorimetry (DSC). These methods, though less glamorous, are by no means less informative. While lacking information on supramolecular structures, the macroscopic properties are extremely useful and often more reliable to determine the phase behavior, including the cmc and the critical micelle temperature (cmt). At high micellar concentrations, the micelles can be close-packed to form a gel. Rheological behavior of such gels, when coupled with structural determinations, could be very interesting and useful in elucidating the nature of such gels. Some data has become available recently.

For convenience, we have listed in Table 3 some important micellar parameters which can be determined by the main physical methods listed in Table 2 and described in previous paragraphs.

TABLE 3 Some Important Micellar Parameters and Conventional Methods of Their Evaluation

	cmc	cmt	n_w	R_g	R_h	r_c	L_s
SLS	+	+	+	(+)		+	
DLS					+		+
SAXS				+			
SANS				+		+	
Surface tension	+						
Dye solubilization	+	+					
Fluorescence	+		+			+	
PGSE-NMR					+		

Three aspects of the physical properties are considered in this review: the dilute region, the phase diagram, and the concentrated region, including the gel formation. The dilute region deals with micelle formation where we are interested in the size, shape, and internal structures of the micelles. It is fairly well understood in terms of the overall qualitative features. More experiments on the chain length (total molecular weight), molecular architecture, and block length ratio dependencies are needed to test the theoretical predictions. The phase diagrams can be complex, but reveal a very rich variety for the Pluronic polyols. Finally, the concentrated region, which is closely connected with our knowledge on the phase diagrams, shows the formation of many kinds of structures which can be detected by scattering techniques.

IV. ASSOCIATION CHARACTERISTICS OF EO/PO POLYMERIC SURFACTANTS IN THE DILUTE REGIME

A. Critical Micelle Concentration and Critical Micelle Temperature

In general, block copolymers form micelles in selective solvents which are thermodynamically good solvents for one type of block but nonsolvents (or poor solvents) for the other type of block. The selfassembly of block copolymer chains in solution can usually be initiated either by an increase in concentration via cmc or by changing the temperature via the cmt. Both cmc and cmt are the fundamental parameters in characterizing the association properties of a given copolymer–solvent system. The cmc is defined as the copolymer concentration, above which the formation of micelles becomes increasingly important. Analogously, the cmt is the transition temperature, above or below which the formation of associated structures becomes appreciable, depending upon whether the selfassembly process is endothermic or exothermic. When dealing with the EPE-type polymeric surfactants in aqueous solution, one of the most distinctive properties is that the selfassembly process is strongly temperature dependent. For a number of Pluronic surfactants, a small increase in temperature, say only 10 °K, may result in a dramatic reduction of the cmc value by a factor of about 10–100. Such a striking temperature dependence is rather unique for this class of polymeric surfactants and constitutes a strong contrast with that of the conventional nonionic surfactants [e.g., $C_m(EO)_n$]. In the latter case, usually a relatively weak temperature dependence is observed. For example, the cmc values of $C_{14}(EO)_8$ at 15 and 40 °C are 6.2×10^{-3} and 4.1×10^{-3} gdm^{-3}, and for $C_{10}(EO)_8$, 7.1×10^{-1} and 3.9×10^{-1} gdm^{-3}, respectively [147], meaning that a temperature increase of 25 °C yields only a 34–45% decrease in the cmc values. For a given system, the cmc and the cmt are interconnected. To a first approximation, the temperature at which the cmc is determined can simply be viewed as the critical micelle temperature for that copolymer concentration. Practically, for Pluronic surfactants there exist two alternative approaches to characterize the micelle formation, namely, either by examining the influence of temperature on a particular physical property at a given concentration as it is often used in light scattering studies, or by following the concentration dependence at a constant temperature as employed in surface tension measurements.

As we have discussed elsewhere [103], it is important to keep in mind that when compared with conventional, low molecular weight surfactants, there is some inherent complexity in the micellization of block copolymers including the Pluronic family. First, the micellization of block copolymers depends strongly on their composition (i.e., the block length ratio), the chain architecture, and the total molecular weight. Some model calculation [149] indicates that the composition inhomogeneity could be appreciable even for a copolymer with a narrow distribution of molecular weight. For commercially available block copolymers, the problem of composition polydispersity becomes even worse, often resulting in a complex state of association and aggregation. Accordingly, no sharp cmc or cmt has been observed for block copolymers. In practice, a certain cmc or cmt range with some notable uncertainty is often detected. Second, as will be seen later, although not in all cases, a noticeable variation or a large difference is often noted between the cmc values reported by different laboratories or by the same laboratory but using different methods. One possible reason is that the copolymer materials examined were from different sources or at least from different batches. More importantly, a variety of techniques can be used to determine the cmc, but their physical principles may be quite different.

Therefore, different methods have different sensitivity to unimers and associated structures, and the nature of the average quantity obtained may be considerably different. For example [109], while light scattering measurements clearly exhibited the transition of unimers into micelles on the concentration curve for a PEP triblock copolymer, Pluronic 17R4 in water at 40 °C, we were unable, under identical conditions, to detect the existence of a cmc in this Pluronic system by using viscosimetry [109]. Similar large discrepancies between different methods, for example, osmometry and light scattering, were observed earlier by Sikora and Tuzar [150] on another class of diblock copolymer–polystyrene-*b*-poly(2-vinylpyridine) in toluene.

The knowledge about the cmc value is of fundamental importance. For common detergents, a large amount of cmc data has been well compiled more than twenty years ago [148]. For water soluble polyoxyethylene/polyoxypropylene (POE/POP) block copolymers, this kind of micellar data has been growing in amount in the past few years but is still scattered in the literature. Therefore, an attempt has been made to compile and to summarize most of the existing cmc data for those EPE and PEP block copolymers which are commercially available. Here again, E and P represent poly(oxyethylene) and poly(oxypropylene), respectively. Considering that the practical demands deal essentially with ambient temperatures, the cmc values at 20, 25, and 30 °C are preferably collected and compiled rather than listing the cmc data over a broad temperature range. Thus, Table 4 gives a list of cmc values, some of which were directly quoted from the relevant references, while others were obtained by interpolation or extrapolation on the basis of existing cmc data at other temperatures. It should be emphasized that in most cases, the latest cmc data are included because they are generally more reliable than those reported in earlier years. Still, because of the composition complexity of block copolymers as discussed above, the reader may note that in Table 4 some variation exists in the cmc values reported by different research groups and in some cases the difference might be quite large. Nevertheless, these cmc data can serve as a useful guide to understand the relationship between the "micellization ability and its temperature dependence," on the one side, and the "composition and structure," on the other side, of the Pluronic polymeric surfactants. Moreover, in the case that the cmc data of interest are absent and thereby some measurements are necessary, it is suggested that the reader should refer to the method of determination that is most appropriate to the real situations in which the problem is raised.

Regarding the association behavior of Pluronic surfactants in aqueous solution, several generalizations can be made as follows:

1. It is generally believed that the dehydration of the POP block with increasing temperature is responsible for the micelle formation [12,13,66]. This explains why Pluronic surfactants show strong temperature dependence in

TABLE 4 Critical Micelle Concentration of Aqueous Solutions of EPE and PEP Block Copolymers

Copolymer	Composition	mw	Temp. (°C)	cmc (gdm^{-3})	Method[a]	Refs.
F-38	$E_{42}P_{16}E_{42}$	4650	20	110[b]	ST	13
			25	100	ST	13
			30	89[c]	ST	13
PF-80	$E_{73}P_{27}E_{73}$	8000	20	82[b]	ST	13
			25	70	ST	13
			30	60	ST	13
PE-6200	$E_6P_{30}E_6$	2650	30	30	ST	13
L-64	$E_{13}P_{30}E_{13}$	2900	20	330[b]	Dye	66
			25	70[c]	Dye	66
			30	15	Dye	66
			20	190[b]	LS	33
			25	44[b]	LS	33
			30	10	LS	33
			20	500[b]	ST	13
			25	110	ST	13
			30	25	ST	13
			20	25[b]	ST	70
			25	11[b]	ST	70
			30	4.7[c]	ST	70
			20	22	Dye	127
P-65	$E_{19}P_{30}E_{19}$	3400	20	410[b]	Dye	66
			25	120[b]	Dye	66
			30	40	Dye	66
			20	410[b]	ST	13
			25	80	ST	13
			30	32	ST	13
F-68	$E_{76}P_{30}E_{76}$	8400	20	290[b]	Dye	66
			25	190[b]	Dye	66
			30	120[c]	Dye	66
			30	230[b]	LS	103
			20	100	F	143
P-84	$E_{19}P_{39}E_{19}$	4200	20	120[c]	Dye	66
			25	26	Dye	66
			30	6	Dye	66

TABLE 4 (Continued)

Copolymer	Composition	mw	Temp. (°C)	cmc (gdm^{-3})	Method[a]	Ref.
P-85	$E_{26}P_{39}E_{26}$	4600	20	200[b]	Dye	66
			25	40	Dye	66
			30	9	Dye	66
			20	16	Dye	127
F-88	$E_{103}P_{39}E_{103}$	11,400	20	190[b]	Dye	66
			25	100[c]	Dye	66
			30	56[c]	Dye	66
			20	6	Dye	127
P-94	$E_{21}P_{47}E_{21}$	4600	40	2×10^{-2}	ST	30
P-103	$E_{17}P_{56}E_{17}$	4950	20	7	Dye	66
			25	7×10^{-1}	Dye	66
			30	1×10^{-1}	Dye	66
P-104	$E_{27}P_{56}E_{27}$	5900	20	20	Dye	66
			25	3	Dye	66
			30	4×10^{-1}	Dye	66
			20	15	ST	13
			25	1.2	ST	13
			30	2.0×10^{-2}	ST	13
P-105	$E_{37}P_{56}E_{37}$	6500	20	22	Dye	66
			25	2.3	Dye	66
			30	2.5×10^{-1}	Dye	66
F-108	$E_{132}P_{56}E_{132}$	14,600	20	130[c]	Dye	66
			25	45	Dye	66
			30	8	Dye	66
L-122	$E_{11}P_{69}E_{11}$	5000	20	2.7[c]	ST	13
			25	1×10^{-1}	ST	13
			30	4×10^{-2}	ST	13
P-123	$E_{20}P_{69}E_{20}$	5750	20	1.8	Dye	66
			25	3×10^{-1}	Dye	66
			30	5×10^{-2}	Dye	66
			20	8.9×10^{-1} [c]	ST	13
			25	4.0×10^{-2}	ST	13
			30	1.9×10^{-2} [c]	ST	13

TABLE 4 (Continued)

Copolymer	Composition	mw	Temp. (°C)	cmc (gdm^{-3})	Method[a]	Ref.
F-127	$E_{99}P_{69}E_{99}$	12,600	20	40	Dye	66
			25	7	Dye	66
			30	1	Dye	66
			20	2.6[c]	ST	13
			25	9.9×10^{-1} [c]	ST	13
			30	3.9×10^{-1} [c]	ST	13
			20	1.4	USV	68
			25	1.0[c]	USV	68
			30	7.7×10^{-1} [c]	USV	68
			30	8×10^{-1}	LS	68
17R4	$P_{14}E_{24}P_{14}$	2650	30	400[b]	LS	109
			40	91	LS	109

[a] Dye, dye solubilization method; LS, light-scattering method; ST, surface tension method; F, fluorescence method; USV, ultrasound velocimetry.
[b] Obtained by extrapolation from the cmt versus reciprocal absolute temperature.
[c] Obtained by interpolation in the same way.

the solution properties. The POP homopolymer with a molar mass of 2×10^3 g mole^{-1} is soluble in cold water (e.g., at temperatures below 10–15 °C) [24], meaning that at low temperatures the Pluronic copolymers exist in the form of unimers even up to high concentrations. Thus, the heat-induced micellization behavior is one of the characteristics of this type of polymeric surfactants. It is well established that in the dilute regime, three temperature regions appear in sequence with increasing temperature, namely, unimer, transition, and micelle regions [13,23,24,36,37,102,103]. The transition region is concentration dependent and shows a coexistence of unimers and micelles. This region usually covers a temperature range of about 10–30 °C, and the equilibrium mixture shifts in favor of micelles with increasing temperature. For most of the commercially available EPE polymeric surfactants, their critical micelle temperatures at which micelles start to form, are between 20 and 50 °C, thus making their usage practical and convenient.

2. For Pluronic copolymers at a given temperature, the cmc decreases exponentially with the POP block length [13,66] (Fig. 1) where the results at 25 °C for normal nonionic surfactants of the polyoxyethylene alkyl ether

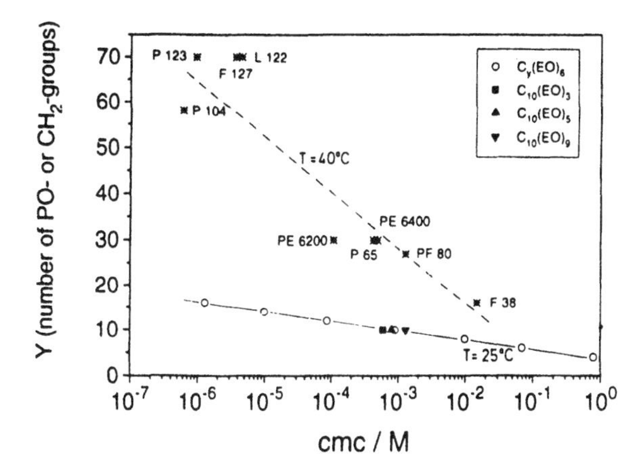

FIG. 1 Logarithm of the cmc values for aqueous solutions of $(EO)_x(PO)_y(EO)_x$ block copolymers with different x-values at 40 °C and for aqueous solutions of alkylpoly (glycol ether) surfactants $C_y(EO)_6$ and $C_{10}(EO)_x$ at 25 °C as a function of the number y of the hydrophobic groups. (*Source*: [13], with permission.)

type are also included. In other words, the number of oxypropylene repeat units plays an important role. Copolymers having a larger hydrophobic POP block form micelles at much lower concentrations. The presence of at least 10–13 PO units in the hydrophobic block is substantial [2,8,13] so that the block copolymer can form micelles at practically accessible concentrations.

3. From the linear relationship between the logarithmic cmc and the number of PO units of the copolymer molecule (see Fig. 1), it is estimated that the transfer energy of a hydrophobic PO group from the aqueous medium into the micellar interior is about 0.2–0.3 kT [12,13] with k being the Boltzmann constant and T the absolute temperature. This value is about 4–5 times less than that of a CH_2 group which shows a value of 1.2 kT. Apparently, the largely reduced hydrophobicity of a PO group (CH_2CHCH_3O), as compared with a CH_2 group, must be attributed to the presence of the polar oxygen atom in the oxypropylene repeat unit.
4. The influence of the POE block length on the micellization is less pronounced than that of the POP block. An increase in the number of EO units leads to a small increase in the cmc and cmt. Thus, the POP block is mainly responsible for the selfassembly process.
5. For a constant POP/POE ratio, the cmc and cmt values decrease with increasing total molecular weight of copolymer surfactants. The lower the relative EO content in the copolymer molecule, the larger is the influence of the total molar mass.

6. From a plot of the logarithmic critical micelle concentration versus the reciprocal of the absolute temperature, a large, positive enthalpy of micellization is derived, demonstrating that for Pluronic polymeric surfactants in an aqueous medium the association process is entropy-driven. This conclusion is also supported by the differential scanning calorimetry results.
7. The chain architecture has a great influence on the micelle-forming ability. As will be discussed in Section IV-F, the selfassembly tendency of the PEP triblock copolymer is, to a large extent, reduced when compared with that of the EPE copolymer of the same composition.

B. Enthalpy of Micellization and Entropy-Driven Nature of the Selfassembly Process

The association behavior of Pluronic copolymers in aqueous medium is strongly temperature dependent. To be specific, the cmc decreases with increasing temperature and the cmt decreases with increasing copolymer concentration. Similar to conventional detergents, micellization of block copolymers in most cases obeys the closed association model, which is characterized by an equilibrium between single copolymer chains (unimers) and polymolecular micelles. This association model requires the existence of a critical micelle concentration which is indeed the case with water soluble polyoxyalkylene copolymer surfactants. In contrast, the open association model is described by a stepwise association and characterized by a distribution of unimers, dimers, trimers, and so on with a series of corresponding equilibrium constants between the respective species. Apparently, the open association mechanism has no connection with the existence of a cmc.

For a closed association mechanism with a relatively large association number and a narrow distribution of association numbers, the standard enthalpy and the standard free energy of micelle formation (the relevant energy changes associated with the transfer of one mole of copolymer from solution to the micellar phase) are given by the relations:

$$\Delta G^0 = RT \ln (X_{cmc}) \tag{1}$$

$$\Delta H^0 = R\left[\frac{d \ln (X_{cmc})}{d(1/T)}\right] \tag{2}$$

where X_{cmc} is the critical micelle concentration in mole fraction units. Assuming that ΔH^0 is approximately a constant within a certain temperature range, Eq. 2 can be integrated to yield

$$\ln (X_{cmc}) = \frac{\Delta H^0}{RT} + \text{constant} \tag{3}$$

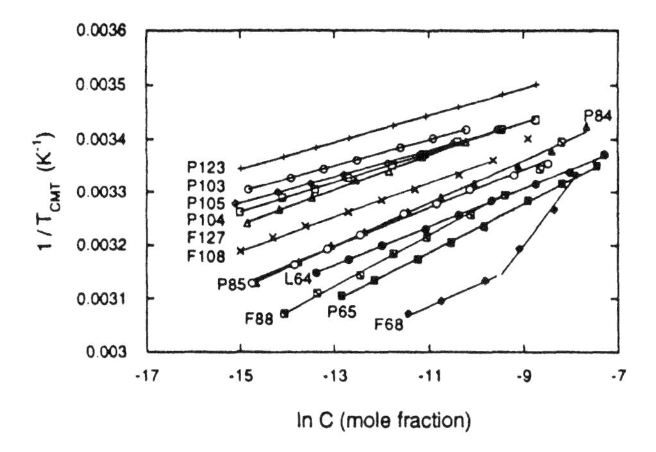

FIG. 2 Reciprocal T_{cmt} versus copolymer concentration plots for various Pluronic surfactants, used for the determination of the micellization enthalpy in terms of the closed association model. (*Source*: from [66], with permission.)

Equivalently, the inverse cmt values can be plotted against the logarithmic copolymer concentration to extract information about the micellization enthalpy. Such plots for a number of Pluronic surfactants, as obtained by Hatton and coworkers [66] based on the dye solubilization data, are shown in Fig. 2. In all cases a good linear fit is obtained except for the Pluronic copolymers F-68, F-88, and F-108 at high concentrations. The authors [66] speculated that because all of these three species contain 80% by wt of POE, the deviations observed might be due to the largely enhanced interactions at high concentrations between the hydrated POE chains belonging to different micellar coronas. Note that for Eqs. 1–3 a basic assumption has implicitly been made, that the association number of micelles is independent of temperature. Although many studies on Pluronic copolymers (e.g., [27,36,68,103]) indicate an increase in association number with increasing temperature, Eq. 3 derived from the closed association mechanism appears quite satisfactory in correlating the cmt data as a function of concentration. The ΔH^0 values thus calculated from the slopes in Fig. 2 are listed in Table 5 under the ΔH^0_{calc} column, where the enthalpy values from other literature sources are also listed, including those of the purified Pluronic L-64 [70] and F-127 [80].

As seen from Table 5, a strongly positive ΔH^0_{calc} value of about 200–400 kJ/mol is obtained for various Pluronic polyols. In the meantime, at the critical micelle temperature for a 1% copolymer solution, the standard free energy of micellization, ΔG^0, has a negative value of about −24 to −28 kJ/mol (Table 6) [66]. Thus, *hydrophobic interactions due to entropy*, although weaker than in

TABLE 5 Enthalpy of Micellization of Pluronic Surfactants in Dilute Solution

Copolymer/ composition		C (wt%)	T_m (°C)[a]	ΔT (°C)[b]	ΔH_{meas} (kJ/mol)[c]	ΔH^0_{calc} (kJ/mol)[d]	Ref.
L-31	$E_1P_{16}E_1$	0.5	51.0		60.6		11
L-35	$E_{11}P_{16}E_{11}$	0.5	73.0		32.4		11
F-38	$E_{42}P_{16}E_{42}$	0.5	81.7		20.6		11
L-42	$E_4P_{21}E_4$	0.5	53.5		87.2		11
L-43	$E_6P_{21}E_6$	0.5	52.1		75.8		11
L-44	$E_{10}P_{21}E_{10}$	0.5	40.6		107		11
PF-20	$E_5P_{27}E_5$	20	23.3	28	96.3		13
PF-40	$E_{12}P_{27}E_{12}$	5	31.4	30	151		13
PF-80	$E_{73}P_{27}E_{73}$	10	40.6	38	106	140	13
L-61	$E_2P_{30}E_2$	0.5	30.4		183		11
L-62	$E_6P_{30}E_6$	0.5	38.9		138		11
		40	23.5	35	74	190	13
L-64	$E_{13}P_{30}E_{13}$	0.5	40.9		185		11
		5	33.8	29	232	200	13
						200[e]	70
						210	109
						230	66
P-65	$E_{19}P_{30}E_{19}$	5	37.4	34	147	180	13
						180	66
F-68	$E_{76}P_{30}E_{76}$	0.5	57.3		91.6		11
		5	57.7	37	123		13
						215	66
P-75	$E_{24}P_{35}E_{24}$	0.5	41.9		143		11
F-77	$E_{52}P_{35}E_{52}$	0.5	50.1		113		11
L-81	$E_3P_{39}E_3$	0.5	28.4		227		11
P-84	$E_{19}P_{39}E_{19}$	0.5	33.4		179		11
						210	66
P-85	$E_{26}P_{39}E_{26}$	0.5	36.2		194		11
						230	66
F-87	$E_{61}P_{39}E_{61}$	0.5	41.7		194		11
F-88	$E_{103}P_{39}E_{103}$	0.5	37.0		166		11
						170	66

TABLE 5 (Continued)

Copolymer/ composition		C (wt%)	T_m (°C)[a]	ΔT (°C)[b]	ΔH_{meas} (kJ/mol)[c]	ΔH^0_{calc} (kJ/mol)[d]	Ref.
L-92	$E_8P_{47}E_8$	0.5	27.1		278		11
L-94	$E_{21}P_{47}E_{21}$	0.5	35.7		210		11
F-98	$E_{123}P_{47}E_{123}$	0.5	43.8		130		11
L-101	$E_5P_{56}E_5$	0.5	20.0		380		11
L-103	$E_{17}P_{56}E_{17}$	0.5	25.3		335		11
						340	66
P-104	$E_{27}P_{56}E_{27}$	5	22.3	24	269	345	13
						295	66
P-105	$E_{37}P_{56}E_{37}$	0.5	27.6		313		11
						330	66
F-108	$E_{132}P_{56}E_{132}$	0.5	31.8		226		11
						265	66
L-121	$E_4P_{69}E_4$	0.5	18.4		401		11
		1	17	24	419		13
		5	15.2	29.5	466		13
L-122	$E_{11}P_{69}E_{11}$	1	18.6	27	310	400	13
P-123	$E_{20}P_{69}E_{20}$	20	13.7	18	351	410	13
						330	66
F-127	$E_{99}P_{69}E_{99}$	0.5	27.4		161		11
		1	23.1	14	485	485	13
		2	21.9	15	495		13
		5	19.6	14	483		13
						255	66
						315[e]	80
17R4	$P_{14}E_{24}P_{14}$					115[e]	109

[a] Temperature at the maximum of the DSC peak.
[b] Width of the DSC peak.
[c] Integrated calorimetric enthalpy in DSC measurements.
[d] Calculated from the slope when ln (cmc) is plotted against $(1/T)$. The last digit has been rounded to 0 or 5.
[e] Purified sample.

TABLE 6 Standard Free Energies (ΔG^0), Enthalpies (ΔH^0), and Entropies (ΔS^0) of Micellization of Various Pluronic Copolymers at Their Critical Micellization Temperature for 1% Solutions[a]

Polymer	mw	N_{PO}/N_{EO}	ΔH^0 kJ/mol	ΔG^0 kJ/mol	ΔS^0 kJ/(mol K)
L-64	2900	1.2	230[b]	–24.5	0.84[c]
P-65	3400	0.79	180	–25.3	0.67
F-68	8400	0.20	215	–28.8	0.76
P-84	4200	1.1	210	–25.2	0.78
P-85	4600	0.75	230	–25.5	0.84
F-88	11,400	0.21	170	–28.5	0.64
P-103	4960	1.7	340	–24.8	1.2
P-104	5900	1.0	295	–25.4	1.1
P-105	6500	0.76	330	–25.6	1.2
F-108	14,600	0.21	265	–28.4	0.98
P-123	5750	1.7	330	–24.9	1.2
F-127	12,600	0.35	250	–27.5	0.94

[a] *Source*: [66].
[b] In this column, the last digit has been rounded to 0 or 5.
[c] The results in this column are shown with two significant figures.

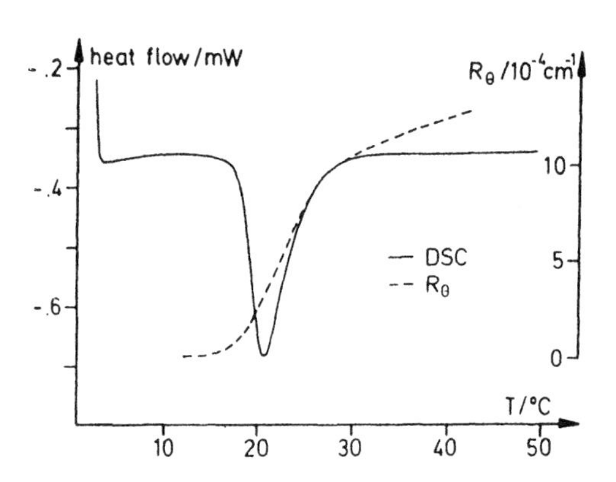

FIG. 3 Rayleigh ratio R_θ from static light scattering measurements (dashed line) and the heat flow from DSC measurements (drawn line) for a 3 wt% aqueous solution of F-127 as a function of temperatures. (*Source*: from [13], with permission.)

the case of low molecular weight surfactants in water, *are responsible for the micelle formation of Pluronic polymeric surfactants in aqueous solution.* This means that the micellization process is entropy driven due to the structural change in water on removal of the hydrocarbon-like PO units.

Differential scanning calorimetry is a direct method to obtain information on the enthalpy of micellization [12,13,78]. The high sensitivity differential scanning calorimetry (HSDSC) permits experiments with a dilute copolymer solution and has been used to collect scanning microcalorimetric transition data for 27 members of the Pluronics family [6–11]. A combination of light-scattering intensity and DSC measurements was employed very recently to monitor the micellization process in a 3 wt% Pluronic F-127 solution [13]. The illustrative results obtained are shown in Fig. 3, where both the heat flow of the calorimeter and the scattered intensity of the copolymer solution are plotted on the same temperature scale. The large endothermic transformation peak occurs in the same temperature region where the scattered intensity shows a profound increase due to micellization. Thus, Fig. 3 clearly demonstrates that the observed DSC signals are an indication of micelle formation. The DSC measurements on different Pluronics show typically a broad endothermic peak which requires three parameters for its characterization: the temperature T_m at the peak maximum, the width ΔT of the peak at its base line, and the integral heat under the peak, yielding, when converted into kJ/mol, the enthalpy of micellization ΔH_{meas}. Note that here the enthalpy change measured from the peak area is not the same as the standard enthalpy of micellization, because not only the interactions between copolymer segments and solvent molecules, but the interactions of segments with each other are also counted in real calorimetric measurements. The T_m and ΔH_{meas} values for various Pluronics obtained by HSDSC using a dilute copolymer solution at $C = 0.5$ gdm^{-3} [11], together with the calorimetric data at higher concentrations obtained by the Bayreuth group [13] are summarized in Table 5.

Several interesting conclusions can be drawn from these DSC and HSDSC results.

1. For the micellization of polyoxyalkylene block copolymers in an aqueous medium, the enthalpy component is significant but unfavorable, which accounts for the remarkable negative temperature dependence of the cmc. In contrast, for block copolymers in organic solvents, a large negative enthalpy of micellization outweighs the opposing entropic factor. Accordingly, a positive slope of the cmc variation against temperature is observed.
2. T_m decreases more or less linearly with the copolymer concentration and also decreases with increasing POP content at a given copolymer concentration. The Pluronic solutions show broad endothermic peaks which are believed to be associated with the broad molecular weight distribution.

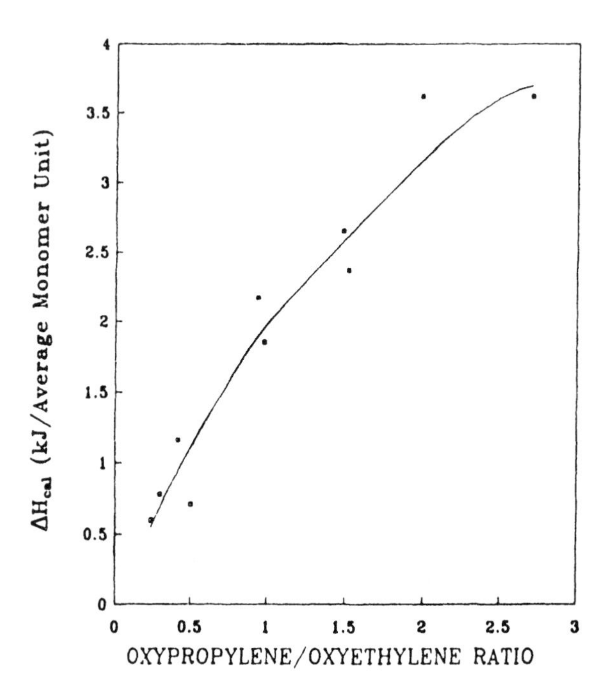

FIG. 4 ΔH_{calc} per average monomer unit versus poly(oxypropylene)/poly(oxyethylene) ratio. (*Source*: from [10], with permission.)

3. The calorimetric enthalpy, ΔH_{meas}, increases with increasing POP content. Importantly, after normalization to the total number of monomer units ($N_{\mathrm{EO}} + N_{\mathrm{PO}}$) in the copolymer molecule, ΔH_{meas} (expressed in kJ/mol per average monomer unit) approaches zero as the POP/POE ratio goes to zero (Fig. 4). This observation indicates that it is the POP part of the copolymer that is responsible for the phase transition (i.e., the micellization process).

Armstrong et al. [10] suggested that the phase transition probably involves the processes of dehydration and association, and perhaps a change in the conformation of the hydrophobe. Such a conclusion is confirmed by the same authors using differential scanning densitometry [10,133] for dilute Pluronic solutions. Later, this kind of analysis showing the dominating contribution of the POP portion was also reported [66] on the basis of the $\Delta H^0_{\mathrm{calc}}$ data per average monomer unit. In addition, it is further argued that the normalized $\Delta H^0_{\mathrm{calc}}$ values seem to be roughly the same for a constant POP/POE ratio regardless of total molecular weight. Further extensive thermodynamic studies on this subject using well-defined copolymer samples would help getting more insights into the association mechanism involved.

C. Association Number, Micelle Size, and Micellar Structure

The static light scattering technique is most often used to determine the molecular weight of the micelle formed in solution. Knowing the molar mass of a single copolymer molecule, it is straightforward to determine the micellar association number which is often called the aggregation number. In our opinion, the term association number appears to be more appropriate because micelles are a thermodynamically stable and equilibrium system, while aggregation usually refers to a nonequilibrium growth of lyophobic colloidal particles into clusters. Another convenient method to determine the association number is fluorescence probing. Both dynamic light scattering and pulsed gradient spin-echo NMR measurements can provide information on the micelle size in terms of hydrodynamic radius R_h. In addition, knowing the association number, n_w, and assuming a nonswelling, liquidlike state for the POP core, one can readily estimate the micellar core radius, r_c, by the relation

$$\frac{4\pi r_c^3}{3} = N_{PO} n_w v_{PO} \tag{4}$$

and

$$v_{PO} = \frac{M_0}{\rho N_A} \tag{5}$$

where v_{PO} and M_0 are the volume and molar mass of oxypropylene monomer, respectively, and ρ is the density of liquid oxypropylene. Here the possible penetration of water and POE into the micellar core is neglected and therefore a lower limit for the core radius is obtained. Furthermore, the numerical difference between R_h and r_c provides a rough estimate on the thickness of the outer corona.

Some association numbers and relevant micellar sizes of Pluronic surfactants are listed in Table 7. Again, the reader may note that the values of association number obtained by different authors for the same Pluronic surfactant are, in some cases, quite different, though the same method of determination (e.g., light scattering) has been used. Presumably, one possible reason is that as mentioned in Sections IV-A and IV-D, for Pluronic surfactants the completion of transition from unimers to micelles covers a temperature range as broad as up to 20–30 °K. This means that in the experiment performed, most likely one is dealing with a mixture of unimers and micelles, and the ratio between them depends upon the experimental conditions chosen for the study. Under such circumstances,

TABLE 7 Weight Average Association Number (n_w), Hydrodynamic Radius (R_h), and Core Radius (r_c), for Various Pluronic Surfactants in Aqueous Solution at Different Temperatures

Copolymer	Medium	Temp. (°C)	n_w	R_h (nm)	r_c (nm)[a]	Method[b]	Ref.
L-64	Water	30	10		1.9	LS	13
		35	24		2.5	LS	13
		40	35		2.9	LS	13
		46	64		3.5	LS	13
		42.5	88	10.2	3.9	LS	102
		47.5	225	12.0	5.4	LS	102
		50.0	356	15.0	6.3	LS	102
		25.9	4			LS	27
		40.0	19			LS	27
		60.0	85			LS	27
		54.1		6.5		PGSE-NMR	27
		40	54			SV[c]	33
		30.0	4			LS	31
	1.0M KCNS	30.0	4			LS	31
	1.0M KBr	30.0	17			LS	31
	1.0M KCl	30.0	29			LS	31
	1.0M KF	30.0	170			LS	31
	0.2M KF	26	6			LS	25
	0.4M KF	26	11			LS	25
	1.0M KF	26	69			LS	25
	0.2M KF	40	26			LS	25
	0.4M KF	40	51			LS	25
	0.5M KF	25.0		2.1		PGSE-NMR	31
	0.5M KF	30.7		4.5		PGSE-NMR	31
	0.5M KF	36.5		7.6		PGSE-NMR	31
	0.5M KF	41.2		15		PGSE-NMR	31
P-65	Water	32	4		1.4	LS	13
		36	11		2.0	LS	13
		40	21		2.4	LS	13

TABLE 7 (Continued)

Copolymer	Medium	Temp. (°C)	n_w	R_h (nm)	r_c (nm)[a]	Method[b]	Ref.
F-68	Water	54.2	19	7.8	2.3	LS	103
		58.9	29	7.9	2.7	LS	103
		68.5	48	7.9	3.2	LS	103
		78.0	68	8.1	3.6	LS	103
	1.0M KF	27	2			LS	29
	1.0M KF	30	4			LS	29
	1.0M KF	40	8			LS	29
	1.0M KF	48	17			LS	29
	1.0M KF	41.0		8.5		LS	29
	1.0M KF	41.0		7.5		PGSE-NMR	29
P-85	Water	25		8.0		LS	17
		40	30	8.0	3.0	LS	17
		50	43	8.0	3.4	LS	17
		30.6		8.5		PGSE-NMR	17
		35.7		9.0		PGSE-NMR	17
	1.0M KBr	30.0	19			LS	31
	1.0M KCl	30.0	25			LS	31
	1.0M KF	30.0	56			LS	31
	1.0M KF	24.5		6.9		PGSE-NMR	31
	1.0M KF	30.6		8.0		PGSE-NMR	31
	1.0M KF	35.7		8.7		PGSE-NMR	31
	1.0M KF	39.4		21		PGSE-NMR	31
P-87	Water	40	14	10.6	2.3	LS	18
		50	14	8.2	2.3	LS	18
F-88	Water	40	17	13.1	2.5	LS	18
		50	17	10.5	2.5	LS	18
P-94	Water	40	36	9.2	3.4	LS	30
P-104	Water	25	32		3.4	LS	13
		30	81		4.7	LS	13
		40	105		5.1	LS	13
		45	121		5.4	LS	13

TABLE 7 (Continued)

Copolymer	Medium	Temp. (°C)	n_w	R_h (nm)	r_c (nm)[a]	Method[b]	Ref.
17R4	Water	40	10	4.0		LS	109
P-123	Water	25	86		5.2	LS	13
		27	211		7.0	LS	13
		35	244		7.3	LS	13
		40	287		7.7	LS	13
		45	297		7.8	LS	13
F-127	Water	20	7		2.2	LS	13
		25	37		3.9	LS	13
		30	67		4.7	LS	13
		35	82		5.1	LS	13
		40	97		5.4	LS	13
		45	106		5.5	LS	13
		10.0	6			LS	68
		30.0	15			LS	68
		35.0	20			LS	68
		40.0	44			LS	68

[a] Calculated from n_w using Eq. 4.
[b] LS, light scattering; SV, sound velocimetry; PGSE-NMR, pulsed-gradient spin-echo nuclear magnetic resonance.
[c] Sedimentation velocity method.

one must be careful in treating the measured data and in making an appropriate extrapolation; otherwise, different final values would be obtained when working with different concentration ranges. Besides, the composition polydispersity of commercial Pluronic copolymers is a factor that always needs to be considered.

As seen from Table 7, the association number and therefore the core radius increases with increasing temperature. Recent small angle neutron scattering (SANS) measurements on aqueous solutions of the Pluronic P-85, F-87, and F-88 [36,37] confirmed that the core radius increases from 2.0 to 6.0 nm in the temperature range 20–100 °C, but appears to be basically independent of concentration below about 30% (w/w). In general, the association number increases with increasing length of the POP block and decreasing length of the POE block. Importantly, upon raising the temperature an increase in association number of the micelle but an approximately constant hydrodynamic radius have been reported for a number of Pluronic polyols, such as F-68 [103], P-85 [17],

P-104 [12], P-123 [12], and F-127 [12,50,68]. This dual effect of the temperature can be understood if we consider the fact that the dehydration of the POE block becomes increasingly important with increasing temperature, resulting in a more or less reduced coronal volume.

It has been well established that in most cases block copolymers form uniform spherical micelles in solution and obey the closed association mechanism with the insoluble block(s) constituting a compact inner core and the soluble block(s) forming a swollen protective corona. For water-soluble Pluronic polyols, the core of the micelle consists of PO groups and the association number of spherical micelles is determined by the length of the POP block. The hydrated POE block constitutes the outer shell with a relatively low density of polymer segments. The micellar core is assumed to be free of water and the core-shell boundary to be fairly sharp. As a whole, such a core-shell model consisting of the dehydrated POP blocks in the core and the hydrated POP blocks at the micellar surface can account for most of the experimental results, at least in the dilute region.

Some model calculations on the Pluronic copolymers [54] suggest that, to a certain degree, the water penetration into the POP core as well as the miscibility of the PO and EO groups may occur. Recent work of small angle neutron scattering on Pluronic L-64 in D_2O [117] also shows the absence of a sharp core-shell boundary in the Pluronic L-64 micelles.

The addition of inorganic salts has a strong influence on the cloud temperatures of Pluronic solutions [31,32]. For example, the Pluronic F-68 solution shows a reduction of 50 °C in its cloud point in the presence of 1.0M KF [29]. While most of the salts reduce the cloud temperature, KCNS and KI allow to increase the cloud temperature. The salts influence the micelle formation and the size and shape of the micelles for the Pluronic solutions. However, when the solution temperature is at the same distance from the cloud point, micelles of about the same size and shape are detected, independent of the type and concentration of the salt added [31]. Table 7 also lists the micellar data of some Pluronic solutions with the salts added.

D. Micellar Shape Transition

For Pluronic P-85 ($E_{27}P_{39}E_{27}$) at low concentrations ($C < 1\%$), a sphere-to-rod transition in the shape of micelles occurs when the temperature is raised to about 70 °C. Brown and coworkers [22] pointed out that above this temperature, both the polarized scattered intensity and solution viscosity increase significantly, and a depolarized scattering component becomes detectable. In addition, the relaxation rate distributions derived from dynamic light scattering measurements exhibit a bimodal character under these conditions. As deduced from the angular dependence, the fast peak portion contains a rotational relaxation component together with the translational diffusion mode. All these observations strongly suggest that elongated micellar structures are formed at high temperatures. Assuming a monodisperse rigid-rod model, the rotational diffusion coefficient deduced from DLS measurements can be used to calculate the rod length. The average rod length thus estimated is about 105 nm at 75 °C in the dilute limit. The rod length is found to increase with increasing temperature. Thus, dynamic light scattering experiments indicate the presence of cylindrical micelles with a length to diameter ratio of about 7 in dilute Pluronic P-85 solutions in the temperature range 70–85 °C [22]. Very recently, a detailed phase diagram for aqueous solutions of Pluronic P-85 was presented based on a combination of various experimental techniques (e.g., small angle X-ray scattering, ultrasonic speed measurements, differential scanning calorimetry, low shear viscosimetry, and light transmission measurements) [23]. This phase diagram of Pluronic P-85 will be given later in Section VI-B (see Fig. 24) when dealing with the gelling properties of Pluronic surfactants. Here we only briefly note that there exists a broad temperature range of about 30 °C for the transition from unimers to spherical micelles. At concentrations below 24 wt% and above this transition temperature range, the spherical micelles grow in size up to about 70 °C. Above this temperature, the region of cylindrical micelles appears, covering a temperature range of about 20–25 °C until the clouding takes place. For another Pluronic copolymer, Pluronic P-94 in water, a change from spherical micelles to rodlike micelles at high temperatures is also proposed [24] to account for the occurrence of a soft gel which is formed at concentrations as low as 1–2 wt%.

For Pluronic P-85 in water, small angle neutron scattering measurements [36,37] show that the core radius of the micelle increases with increasing temperature. Close to 70 °C, the core radius has a value of about 5.0 nm and therefore corresponds, on average, to a length of 0.25 nm for each PO unit. This means that around this temperature the POP blocks in the micelle core are fully extended. Accordingly, a further increase in the micellar association number as caused by raising the temperature must lead to a marked deviation of the micellar shape from spherical symmetry. Based on the same arguments, for those Pluronic copolymers in which the hydrophobic part is almost twice as large as the hydrophilic part, such as Pluronic P-104 and P-123, the temperature-induced micellar shape transition could also be expected [13]. Nevertheless, further studies to ascertain the temperature-induced micellar shape changes should be worthwhile.

Model calculations of phase diagrams using a mean field lattice theory for multicomponent mixtures have recently been made for the Pluronic copolymers P-95, P-104, and P-105 in water [54]. By allowing the EO and PO segments of the copolymer to adopt different states that depend on the temperature and the solution composition, these model calculations are able to predict the regions for the unimer solution, spherical micelles, and elongated rods as well as the two-phase region as a function of temperature and concentration. In general, these calculations give a qualitatively correct description about the changes in the location of different phase regions when varying the composition of the Pluronic copolymers.

E. Composition Inhomogeneity and Anomalous Micellization

In the micellization of block copolymers in selective solvents, a peculiar phenomenon, sometimes referred to as *anomalous micellization* has been reported by a number of research groups on a variety of block copolymer systems [3]. In most cases, this anomaly is clearly observed when dealing with the temperature-induced micellization behavior and in some cases can be detected by examining the concentration dependence via the critical micelle concentration. The anomalous behavior manifests itself essentially as the presence of large particles before the onset of micellization, thus exhibiting a strong opalescent appearance. This opalescence disappears either by varying the temperature or by further addition of copolymer. Several speculative mechanisms have been suggested, such as the formation of anisotropic ellipsoids through the linear association of monomolecular micelles, coexistence of metastable worm-like structures with regular spherical micelles, and the transient formation of emulsion droplets [see 3,102]. Due to a lack of direct, strong experimental evidence, however, the origin of such a strange phenomenon associated with block copolymers in solution has been a puzzle for some time.

For polyoxyalkylene block copolymers, the anomalous micellization behavior has most clearly been observed on aqueous solutions of Pluronic L-64 [101,102,168]. The anomalous region exists in a relatively narrow range between the two temperature intervals corresponding to the unimer and micelle regions, respectively. In these two regions, the measurements on the physical properties of interest are well reproducible. But, in the anomalous middle region where a very strong scattering peak is monitored, the scattered light intensity and the intensity-intensity autocorrelation function measured are often less reproducible, indicating that most likely, the large particles or aggregates detected are not thermodynamically equilibrium structures. A careful CONTIN analysis [151,152] performed in this anomalous region indicates the coexistence of either two species (unimers and large particles) or three species (unimers, micelles, and large particles) [17,115]. Although the scattering by large particles is dominating, the large particle portion is estimated to constitute only a few thousandths to at most a few percent of the total weight; this means that single copolymer chains are still quantitatively overwhelming. Importantly, the location of the anomalous scattering region is concentration dependent, shifting to a high temperature with decreasing Pluronic L-64 concentration. This is exactly in parallel with the critical micelle temperature of the Pluronic L-64 solution which also moves to a high value with decreasing concentration.

One of the inherent characteristics of block copolymers is their complexity in composition (i.e., composition polydispersity). As mentioned above, the chemical heterogeneity could be appreciable even for a copolymer with a narrow molar mass distribution. It is known, for example, that triblock copoly(oxyethylene/oxypropylene/oxyethylene)s are usually contaminated by small quantities of more hydrophobic diblock copolymers that appear as by-products arising from the transfer reaction in the preparation process [73]. Based on this concept the anomalous micellization behavior can be readily understood in the following way. First, it is known that the anomalous behavior is observed in an intermediate region between the two temperature regimes where unimers and micelles exist, respectively. Next, the composition polydispersity of a block copolymer means a variation in the amphiphilic properties among individual copolymer chains. A small portion of the copolymer specimen may exist with a higher content of the insoluble block. Such minor components would become insoluble, by varying the solution temperature or by altering the composition of the mixed solvent, before the onset of micellization of the major component is reached. In other words, this minor "insoluble" portion could be responsible for the anomalous behavior. Let us take the heat-induced micellization as an example. There are two important parameters: 1) the phase separation temperature of the minor component, and 2) the critical micellization temperature of the central portion of the system studied. When the temperature is low enough, all the copolymer chains are soluble (i.e., molecularly dispersed). Upon raising the temperature, we first meet the phase separation of the minor less-soluble components, which leads to the formation of a dilute dispersion of large particles stabilized by the adsorbed layer of the major component, resulting in anomalous scattering. Later, as the cmt of the major component is reached, the insoluble minor components can either be incorporated into the micelle core or form mixed micelles. Consequently, the scattering peak region is followed by a micelle region showing normal light scattering. If the above mechanism is correct, the anomalous behavior would disappear provided that the minor components which are insoluble in the peak region, have been removed. The same principle also applies to the interpretation of anomalous scattering induced by varying the solvent composition when using a mixed solvent.

One may also claim that the anomalous micellization behavior could be an inherent property of block copolymers. For instance, it could be attributed to correlated concentration fluctuations in the transition region [69] or to the existence of large, secondary associated structures which are in equilibrium with small micelles. To distinguish between these possible mechanisms, a filtration experiment was carried out in the peak region in order to remove the minor insoluble components. The details have been described elsewhere [101,102].

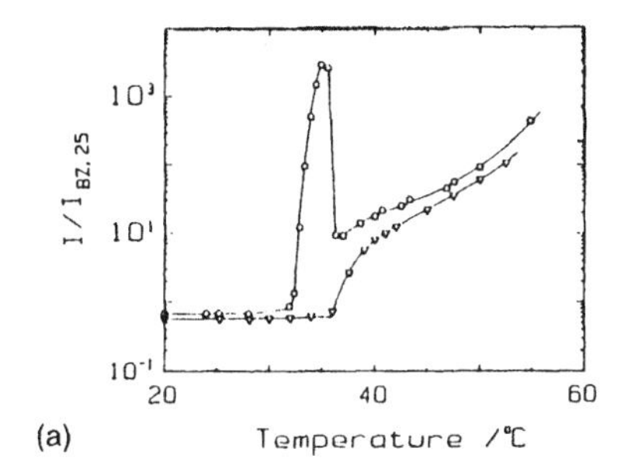

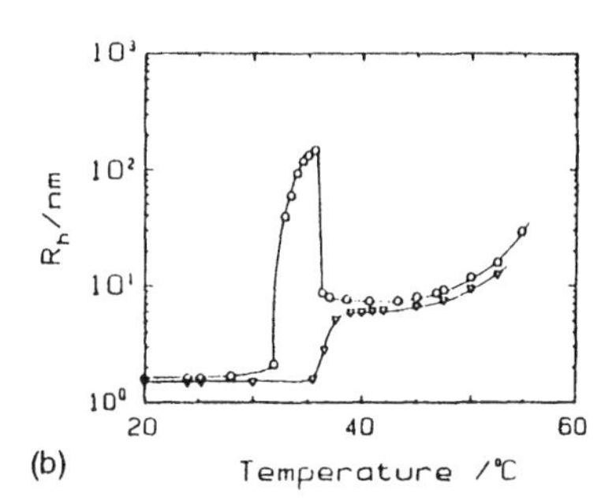

FIG. 5 (a) Effect of filtration on the temperature dependence of scattered light intensity relative to that of benzene at 25 °C for Pluronic L-64 in water: circles denote before filtration with solution concentration C = 10.0 mg/mL; inverted triangles denote after filtration with solution concentration C = 9.65 mg/mL. (b) Effect of filtration on the temperature dependence of apparent hydrodynamic radius. Same symbols as in (a). (*Source*: from [101], with permission.)

The idea is that if the anomalous behavior were due to either local concentration fluctuations in the transition region or large elongated particles in equilibrium with normal, spherical micelles, filtration itself could not eliminate it, because anomalous scattering would reappear in the filtrate after the equilibrium is restored on standing. But, if the composition heterogeneity mechanism were correct, a normal micellization behavior should be observed on the temperature curve after an effective filtration.

Experimentally, the above arguments have been tested and confirmed for both EPE and PEP block copolymers, namely, Pluronic L-64 ($E_{13}P_{30}E_{13}$) [101,102], 10R5 ($P_8E_{22}P_8$) [110], and 17R4 ($P_{14}E_{24}P_{14}$) [109] in water. Figures 5a and 5b obtained on L-64 in water clearly demonstrate that after effectively removing the large particles by filtration at the temperature corresponding to the maximum of scattered intensity, the anomalous region vanished completely and a normal transition was observed. Also, as evidenced by NMR spectroscopy [102], the small amount of residue on the filter indeed showed a higher content of hydrophobic oxypropylene in comparison with the average composition of the sample. In summary, the existence of anomaly requires: 1) composition heterogeneity, and 2) the phase separation of the anomaly-causing fraction occurs before the onset of micellization of the major component. The second point explains why only some block copolymers in a selective solvent exhibit anomalous behavior.

From their fluorescence probing results, Almgren et al. [25] concluded that Pluronic L-64 is polydisperse in composition and that about 2% of the material is more hydrophobic than the major component, thus forming aggregates at lower temperatures and with less amount of potassium fluoride added than the bulk material. The HPLC analysis of the copolymer sample exhibited a bimodal elution profile, suggesting the presence of about 3% of diblock contaminant. Similar degrees of composition contamination (up to 6 wt%) were also reported on Pluronic P-85 [23], P-94 [24], and F-127 [50]. Light scattering measurements on Pluronic L-64 [27], P-85 [17], and P-94 [30] in aqueous solution clearly indicate the presence of a small amount of large clusters or aggregates at low temperatures together with the dominating amount of unimers, but these aggregates disappear at elevated temperatures where micelles become appreciable in quantity. All these observations are consistent with the above discussed composition heterogeneity mechanism of the anomalous micellization behavior.

Some surface and solution properties, such as surface tension, cloud point and light-scattering intensity, are quite sensitive to "impurities," the existence of which often disturb the results of measurements. Booth and coworkers reported the removal of the more hydrophobic impurities from Pluronic L-64 by extraction with hexane [70] and from Pluronic F-127 by fractionation with a solvent mixture [80], as verified by the elimination of the existing surface tension minimum.

F. Effect of Chain Architecture

The chain architecture of block copolymers is expected to have influence on their selfassembly behavior in both dilute and concentrated regimes. For an AB diblock copolymer in a solvent that is selective for either A or B as well as an ABA triblock copolymer in a selective solvent for the terminal blocks, the association into micelles is relatively straightforward, obeying the closed association model and yielding the core-shell structure composed of a relatively compact micellar core and a swollen protective corona. However, for the same type of triblock copolymers but in a solvent selective toward the central block, a variety of possible associated structures could be expected. Flower-like micelles would form in the case that the two end blocks constitute a part of the same micellar core whereas the central block takes a loop conformation. The extended soluble central block may function as a bridge to connect the small clusters composed of insoluble end blocks, thus leading to a branched structure at low concentrations or a gel-like network at high concentrations. Besides, an intermediate pattern is also possible in that some of the coronal blocks show a looping geometry, but the other copolymer chains may have one of their end blocks located in the core and the other dangling in solution. Accordingly, there should be no simple answer if one asks what kind of association structure forms for a triblock copolymer when dissolved in a selective solvent for the middle block. It all depends on the resulting balance between two competing factors: 1) an entropy loss due to the loop conformation of the central block would favor the formation of branched structures, and 2) an interfacial energy penalty that accompanies the transfer of the poorly solvated block from the core into solution would favor the flower-like micellar structure. In essence, the lengths of the two constituent blocks and the interfacial energy between the core and the solvent are both of great importance. Some recent work of computer simulation [125,126] on short triblock copolymers supports the above analysis.

While a large number of studies have been devoted to the micelle formation and structural characteristics of polymeric micelles for the EPE-type Pluronic surfactants, little is known about the association properties of the PEP-type polymeric surfactants (the BASF trade name is Pluronic R). Very recently, a detailed comparison of association properties between the EPE and PEP block copolymers was made to study the effect of chain architecture [109]. Pluronic L-64 ($E_{13}P_{30}E_{13}$) and 17R4 ($P_{14}E_{24}P_{14}$) were chosen for this purpose. It should be noted that these two block copolymers have the same ethylene oxide content (40% by weight) and differ only slightly in the total molecular weight. Therefore, they constitute a good pair of subjects for comparison. Pluronic 17R4 was purified so that the anomalous scattering was removed. Figures 6a and 6b show the phase diagrams of Pluronic 17R4 and L-64 in water, where regions I, II, and III are the one-phase unimer region, one-phase micelle region, and two-phase region of two immiscible isotropic solutions, respectively. Several important points are seen from Figs. 6a and 6b:

1. The phase behavior of Pluronic L-64 in the concentration range studied is simple. Pluronic L-64 forms micelles over a large temperature range as well as a broad concentration range (region II in Fig. 6b). Unimers exist only in a small corner in the T-C plane (i.e., at low temperatures and in dilute solutions).
2. In contrast to Pluronic L-64, the 17R4 unimer region extends over a broad concentration range and a large temperature range. Pluronic 17R4 forms flower-like micelles only in a largely restricted area (i.e., at high concentrations within a narrow, wedge-shaped temperature region, the width of which expands moderately with increasing concentration). Importantly, when raising the temperature in the dilute region ($C < 75$ gdm^{-3}), the cloud-point curve rather than the cmt curve is encountered first and therefore the formation of polymolecular micelles becomes practically impossible, no matter whatever temperature has been selected. Additionally, aqueous solutions of purified Pluronic 17R4 show higher cloud points as compared with the unpurified sample.
3. For Pluronic 17R4 in water, light scattering measurements clearly indicate the existence of a cmc which decreases with increasing temperature until the cloud point is reached. The small region II consists of an equilibrium mixture of unimers and micelles, with the equilibrium shifting in favor of micelles when the concentration is increased. Both the unimer and micelle portions show a relatively narrow size distribution as evidenced by the CONTIN analysis of dynamic light scattering results [109].

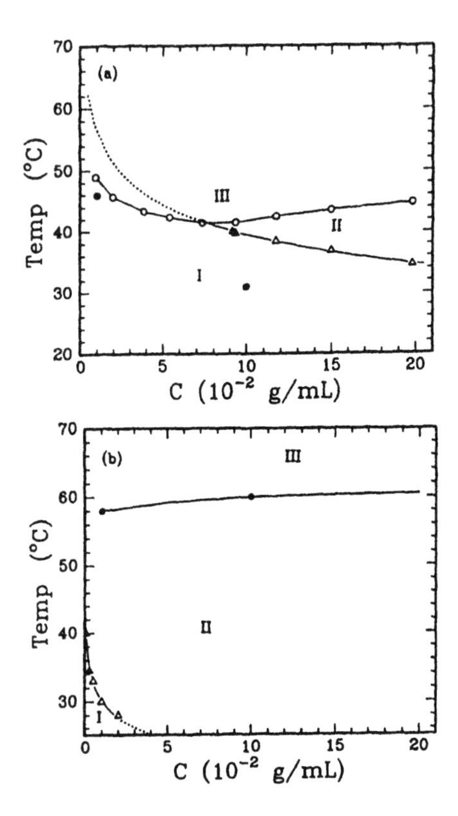

FIG. 6 (a) Phase diagram of 17R4 in water. Regions I, II, and III are the one-phase unimer region, one-phase micelle region, and two-phase region of two immiscible isotropic solutions, respectively. (b) Phase diagram of L-64 in water. Also, three phase regions are shown. (*Source*: from [109], with permission.)

To summarize the effect of chain architecture, Table 8 [109] gives a compari son of micellar parameters between Pluronic 17R4 and L-64 micelles in water, including the thermodynamic functions of micellization. The entropy penalty associated with the looping geometry of the middle block does not preclude the possibility of micelle formation, but largely reduces the tendency toward self-assembly. For instance, at 40 °C, Pluronic 17R4 has a cmc value 100 times larger than Pluronic L-64 and the association number is 9 times less. Regarding the micellar structure, the magnitudes of the two competing factors—the entropy loss due to the backfolding of the POE block and the interfacial energy term of the insoluble POP block against water—may be estimated. Following ten Brinke and Hadziioannou [153], the entropy penalty due to the loop formation is given by

TABLE 8 Comparison in Micellar Parameters Between 17R4 and L-64 Micelles in Water[a]

	Pluronic L-64	Pluronic R 17R4
Composition	$(EO)_{13}(PO)_{30}(EO)_{13}$	$(PO)_{14}(EO)_{24}(PO)_{14}$
HLB value	15	7–12
cmc at 40 °C, g/mL	9.0×10^{-4} [b]	9.1×10^{-2}
M_w, g/mol	2.8×10^5 (42.5 °C)[c]	2.65×10^4 (40.0 °C)
Association no. (n_w)	88 (42.5 °C)[c]	10 (40.0 °C)
A_2, cm^3 mol g^{-2}	~6.0×10^{-4} (42.5 °C) [c]	~6.0×10^{-5} (40.0 °C)
R_h, nm	10.2 (42.5 °C)[c]	4.0 (40.0 °C)
ΔH^0, kJ/mol	210	115
ΔG^0, at 40 °C, kJ/mol	–21	–9
ΔS^0, at 40 °C, kJ/(mol/K)	0.74	0.40

[a] *Source*: [109].
[b] *Source*: [70].
[c] *Source*: [102].

$$\frac{G_{\text{loop}}}{kT} = \left(\frac{3}{2}\right) \beta \ln N_{\text{EO}} \tag{6}$$

where N_{EO} (= 24) is the number of the EO units in the central block. The coefficient β that depends on the lengths of the two constituent blocks, has a limiting value of 0.3–0.4 [126]. The free energy term associated with the interface formation between the insoluble core and the solvent is approximately expressed by the relation

$$\frac{G_{\text{int}}}{kT} = \pi\, a^2\, N_{\text{PO}}^{2/3}\, \sigma \tag{7}$$

where a is the size of the oxypropylene unit and estimated approximately to be 0.56 nm, N_{PO} is the number of the PO units in the terminal block, and σ is the relevant interfacial tension (in kT unit). The POP/water interfacial tension is not readily available, but could be roughly estimated to be 4–8 mN/m [154]. Thus, one finally has G_{loop} = 1.9 kT and G_{int} = 6–11 kT. This kind of evaluation has a certain amount of crudeness, but accounts for the formation of flower-like micelles in the case of Pluronic 17R4 in water, as shown experimentally. Recent studies [111,115] on a long chain triblock copolymer, poly(*tert*-butylstyrene)-polystyrene-poly(*tert*-butylstyrene), in a selective solvent for the polystyrene block provide similar conclusions about the influence of the sequence of blocks.

On the other hand, a diblock EP and a triblock EPE of the same overall composition are expected to show similar micellar properties in water. A recent comparative study on $E_{26}P_{29}$ and $E_{14}P_{30}E_{14}$ [77] which were synthesized for this purpose, has demonstrated that at a temperature of 50 °C, the values of surface and micellar properties determined, such as the surface activity, cmc, n_w and R_h, are virtually the same within the uncertainties of their evaluation, except for the area at the air/water interface occupied per copolymer molecule at the cmc.

G. Surface Activity and Surface Adsorption

Like low molecular weight surfactants, micellization and adsorption are the two fundamental aspects of surface-active block copolymers, reflecting their bulk and interfacial properties, respectively. In this section we will briefly discuss the surface properties of Pluronic polymeric surfactants, focusing on the surface activity (the ability to reduce the surface tension of water) and the surface adsorption (the concentrating process of the solute at the air-water surface). Because of the limited space, the topic on the adsorption behavior of Pluronic copolymers at the solid/liquid interface (e.g., [46,47,97,128,129, 134–136,138]) which is closely related to the steric stabilization of dispersions, will not be discussed in this chapter.

In general, the surface tension of a surfactant solution decreases monotonically until the cmc is reached and then stays constant above the cmc. The Gibbs adsorption equation which applies to all adsorption processes has its general form

$$d\sigma = -\sum \Gamma_i d\mu_i \tag{8}$$

where $d\sigma$ is the change in the solvent surface tension, Γ_i the surface (excess) concentration of the component i of the system, and $d\mu_i$ the change in chemical potential of the component i. For a single nonionic surfactant solution, this equation can be written in the following simple form when applied to the air/water interface,

$$d\sigma = -\Gamma RT d \ln C \tag{9}$$

where Γ is the surface (excess) concentration of surfactant molecules adsorbed at the surface and C is the bulk concentration of the surfactant. The surface area per surfactant molecule, A, is calculated by the relation

$$A = \frac{1}{\Gamma N_A}$$

with N_A being Avogadro's number. Thus, surface tension measurements provide a simple and convenient way to determine the adsorbed amount as well as the surface area per surfactant molecule at the air/water interface against the bulk concentration, and thereby provide a means to construct the surface adsorption isotherm. For concentrations below but near the cmc, the slope of the σ versus log C curve is essentially constant, meaning that the adsorbed amount of the surfactant has reached a constant maximum value. In other words, in this region the surface is saturated with surfactant (i.e., at a complete coverage). Under these conditions, several characteristic physical quantities can be determined. They are: 1) the cmc; 2) the surface tension at the cmc, σ_{cmc}; 3) the cmc/C_{20} ratio with C_{20} being the copolymer concentration at which the solvent surface tension is lowered by 20 mN/m (this ratio reflects the competition between micellization and adsorption); and 4) the surface concentration at a complete coverage, Γ_{sat}, or the surface area per copolymer (surfactant) molecule, A, at the cmc.

In the case of Pluronic copolymers in aqueous solution, as will be seen later, a complexity of the surface tension profile has often been observed, particularly when a wide range of concentrations, say from 10^{-5}% to 10%, or a broad temperature range is covered. Correspondingly, the values of these characteristic parameters derived depend largely on which break on the surface tension curve has been selected for consideration.

Prasad et al. [127] made extensive surface tension measurements on seven Pluronic copolymers at 20 °C, covering a very broad concentration range, from 10^{-7}% to 10%. They observed the existence of two breaks on the surface tension curves. In all cases, however, up to a concentration of about 10%, no plateau region of the surface tension was observed. The inflection points that occurred in the vicinity of 1×10^{-3}% concentration were, as they speculated, due to the formation of monomolecular micelles where the conformational change of the single copolymer chain may lead to a close-packed entity with the hydrophobic block coiled in the interior and shielded by the hydrophilic EO units. These surface tension break points were at concentrations much lower than the inflection points they observed based on the solubilization data of benzopurpurine and iodine by the same copolymer solutions. For the Pluronics L-42, L-44, L-62, L-64, F-68, P-85, and F-88, the values of the surface concentration and the area per surfactant molecule were derived from the break in surface tension around $C = 1 \times 10^{-3}$% (Table 9). Very recently, Hoffmann and coworkers [13] reported their systematic study on various Pluronic surfactants, including surface tension measurements at different temperatures. They found that the temperature has a significant influence on the shape of the surface tension curve, as clearly shown for Pluronic P-104 in water in Fig. 7. It can be seen that while both the cmc and the slope [= $d\sigma/d$ (log C)] at the cmc strongly depend upon the temperature, the surface tension value at the cmc shows a little temperature dependence.

TABLE 9 Surface Tension at the cmc (σ_{cmc}), the cmc/C_{20} Ratio, and Surface Concentration (Γ_{sat}) and Surface Area per Copolymer Molecule (A) at a Complete Coverage for Various Pluronic Surfactants in Aqueous Solution

Copolymer	Temp. (°C)	σ_{cmc} (mN/m)	cmc/C_{20} [e]	Γ_{sat} (mol/m^2)	A (Å^2/molecule)	Ref.
F-38	20			7.44×10^{-7}	225[d]	13[a]
	25	39.7		7.97×10^{-7}	210	13
	40	37.5		8.8×10^{-7}	190	13
L-42	20			1.48×10^{-6}	120	127[b]
L-44	20			1.14×10^{-6}	145	127
L-62	20			2.6×10^{-6}	65	127
PF-80	15			5.75×10^{-7}	290	13
	25	41.4		6.5×10^{-7}	255	13
	30	39.5		6.78×10^{-7}	245	13
	45	38.4		7.14×10^{-7}	230	13
PE-6200	15			5.34×10^{-7}	310	13
	20			6.7×10^{-7}	250	13
	25	37.8		7.4×10^{-7}	225	13
	30	37.3		7.7×10^{-7}	215	13
	40	36.4		1.15×10^{-6}	145	13
L-64	15			5.04×10^{-7}	330	13
	20			5.86×10^{-7}	285	13
	25	37.8		6.27×10^{-7}	265	13
	30	37.3		6.28×10^{-7}	265	13
	40	36.3		9.31×10^{-7}	180	13
	20			1.54×10^{-6}	110	127
P-65	15			5.3×10^{-7}	315	13
	25	37.8		6.0×10^{-7}	280	13
	30	37.0		6.67×10^{-7}	250	13
	40	36.9		8.8×10^{-7}	190	13
	25		40,000	1.41×10^{-6}	120	67[c]
	35		4000	1.57×10^{-6}	105	67
F-68	20			1.15×10^{-6}	145	127
P-85	25		25,000	1.36×10^{-6}	120	67
	35		2000	1.58×10^{-6}	105	67
	20			2.6×10^{-6}	65	127

TABLE 9 (Continued)

Copolymer	Temp. (°C)	σ_{cmc} (mN/m)	cmc/C_{20} [e]	Γ_{sat} (mol/m^2)	A (Å^2/ molecule)	Ref.
F-88	20			1.51×10^{-6}	110	127
P-103	25		500	2.7×10^{-6}	60	67
	35		20	3.2×10^{-6}	50	67
P-104	15	38.2		3.86×10^{-7}	430	13
	20	37.8		4.74×10^{-7}	350	13
	25	37.1		5.32×10^{-7}	310	13
	30	36.9		1.15×10^{-7}	145	13
	40	36.5		2.66×10^{-6}	60	13
	25		1500	2.2×10^{-6}	75	67
	35		80	2.5×10^{-6}	65	67
P-105	25		2000	1.7×10^{-6}	100	67
	35		100	2.0×10^{-6}	85	67
F-108	25		2000	1.09×10^{-6}	150	67
	35		200	1.23×10^{-6}	135	67
L-122	15	37.6		3.8×10^{-7}	435	13
	25	36.5		1.07×10^{-6}	155	13
	30	36.4		1.38×10^{-6}	120	13
	40	36.4		1.59×10^{-6}	105	13
P-123	17	36.5		4.63×10^{-7}	360	13
	25	36.3		8.26×10^{-7}	200	13
	35	36.3		1.23×10^{-6}	135	13
	43	36.1		4.57×10^{-6}	35	13
	25		125	3.3×10^{-6}	50	67
	35		20	4.2×10^{-6}	40	67
F-127	15	40.6		4.53×10^{-7}	365	13
	42	40.3		1.39×10^{-6}	120	13

[a] In [13], the Γ_{sat} and A values refer to the second break.
[b] In [127], the Γ_{sat} and A values refer to the break occurring at $C = 1 \times 10^{-3}$%.
[c] In [67], the Γ_{sat} and A values refer to the first break.
[d] In this column, the last digit has been rounded to 0 or 5.
[e] For definition of C_{20}, see Section IV-G (p. 103).

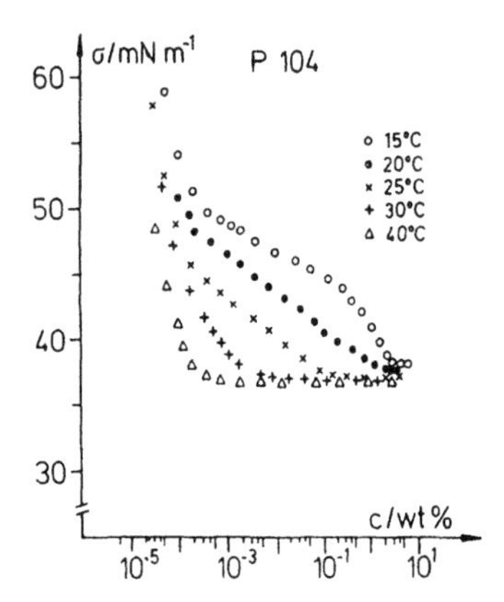

FIG. 7 Surface tension σ for aqueous solutions of P-104 as a function of concentration at various temperatures. (*Source*: from [13], with permission.)

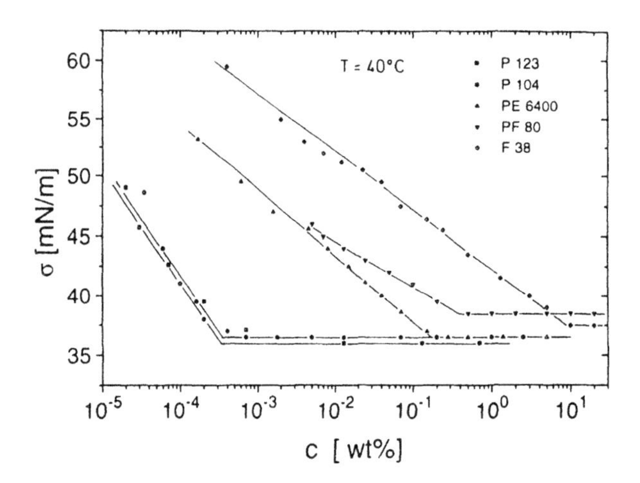

FIG. 8 Surface tension σ for various block copolymer solutions as a function of the logarithm of concentration at 40 °C. (*Source*: from [13], with permission.)

Interestingly, at a high temperature of 40 °C (Fig. 8), all five Pluronic polyols had a concentration region where the surface tension of the solution decreased linearly with the logarithmic concentration until the cmc is reached and the surface tension remained constant beyond that concentration. Thus, at this temperature, the concentration dependence of the surface-active behavior becomes as simple as that of the conventional detergent. The surface activity and surface adsorption results they obtained on various Pluronic polyols at different temperatures are listed in Table 9. Almost in the same period, another extensive study on the surface activity of Pluronic copolymers was presented by the MIT group [67], focusing on the significance of the two breaks occurring on the concentration curve of the surface tension at two different temperatures 25 and 35 °C (Fig. 9). For comparison, they used Pluronic copolymers having the same size POP block but different POE block length as well as Pluronic surfactants showing the same POE/POP ratio but different total molecular weight. They pointed out that neither polydispersity nor impurities should be responsible for the first break occurring around $C = 1 \times 10^{-3}\%$ on the surface tension curve. They also suggested that the origin of this first break arises from the rearrangement of the Pluronic molecules in the adsorbed layer into a more compact structure at a complete coverage on the air/water interface. They claimed that the second break occurring at a higher concentration as the onset of the plateau region corresponds to the so called cmc and signifies the formation of polymolecular micelles. Note that the surface tension values for various Pluronic polyols at the first break vary between 40 to 50 mN/m, while the values at the second break are about 35–40 mN/m. Importantly, the cmc values thus determined are in agreement with those they obtained from dye solubilization experiments performed on the same copolymer samples. Their results are included in Table 9. On the other hand, the Bayreuth group [14] recently stated that the occurrence of the two breaks (at $C \approx 1 \times 10^{-4}$ and 7×10^{-1} wt%, respectively) on the surface tension curve of Pluronic F-127 is likely the result of the broad molecular distribution of the copolymer. While the most surface-active portion starts to associate in the solution at the first break point, the less surface-active part in the unimer form increases with increasing bulk concentration until the second break is reached where all the material forms micelles. The Manchester group [80] reported a profound minimum in surface tension for unpurified Pluronic F-127 in water.

Table 9 gives a summary of the values of those characteristic parameters which are in connection with the surface activity and surface adsorption for various Pluronic surfactants at different temperatures. Note that a large variation occurs in the listed values of the saturated adsorbed amount and the surface area per copolymer molecule. This is mainly because different research groups paid attention to different break points on the surface tension curve. Moreover, due to the low diffusion rate to the surface in the low concentration range, a

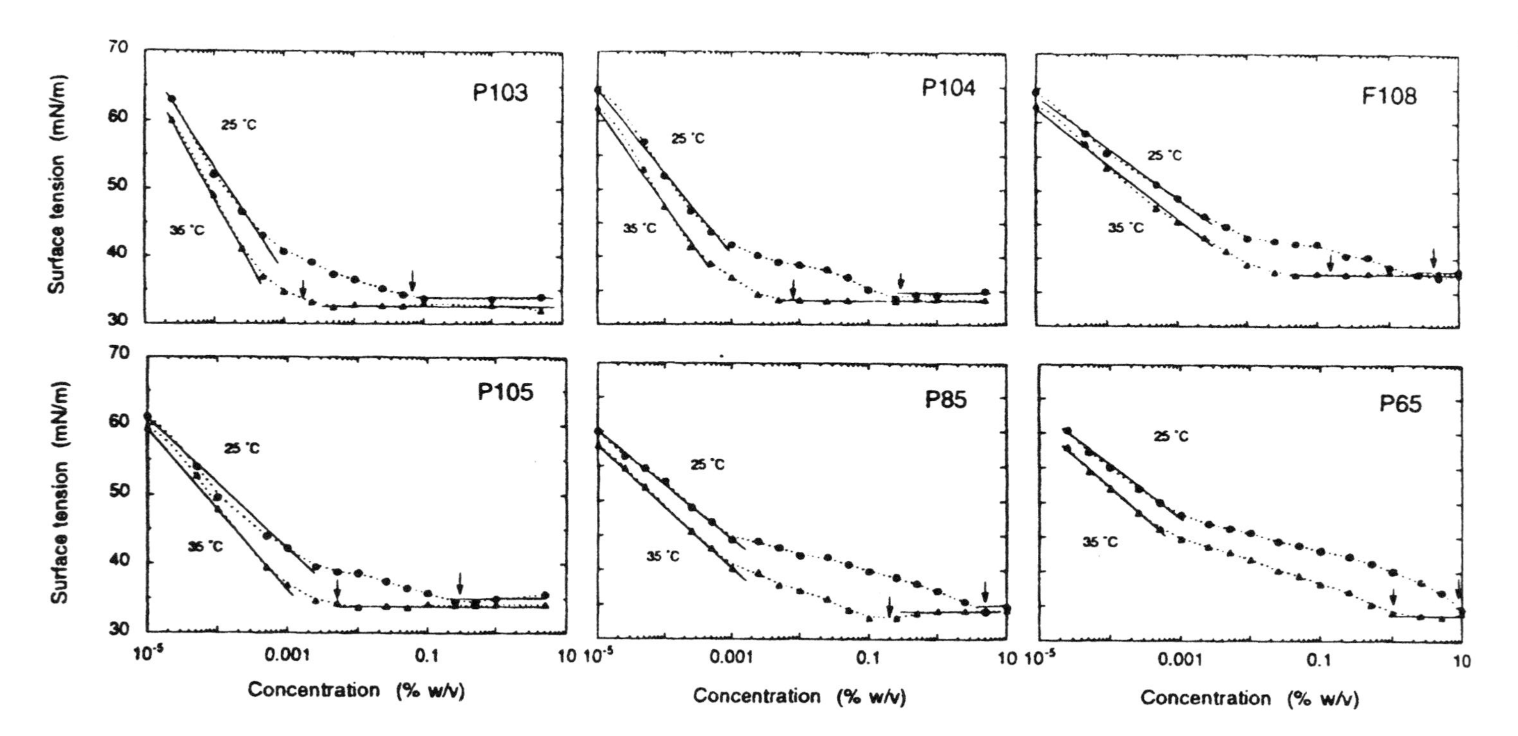

FIG. 9 Surface tension data for the Pluronic solutions, plotted as a function of concentration at two temperatures (25 and 35 °C). The arrows mark the cmc obtained for the same copolymer using dye solubilization. (*Source:* from [67], with permission.)

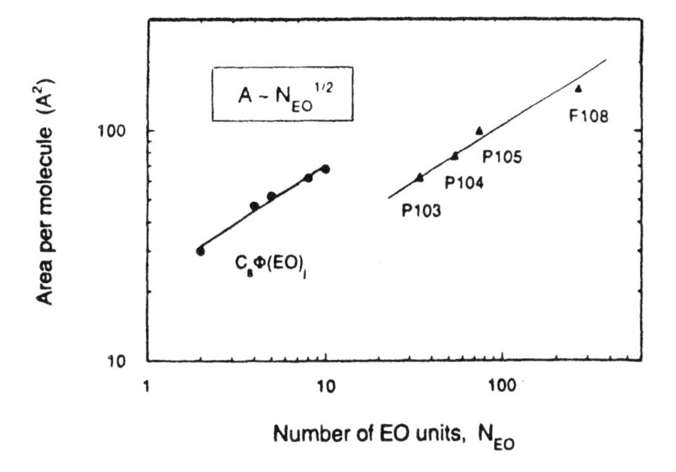

FIG. 10 Area per molecule as a function of the number of EO units in the copolymer molecule, for the Pluronic family with the same POP and increasing POE length. Also shown in the plot are areas per molecule for a series of [(octylphenoxy)-ethoxy]ethanol surfactants. (*Source*: from [67], with permission.)

sufficiently long waiting time (sometimes hours or more) is often required to ensure an equilibrium value for the surface tension obtained in the dilute regime [71,77]. The uncertainty in the slope evaluation of the surface tension curve, particularly obtained at low temperatures, and the composition variation of block copolymers from different manufacturers or from different batches are the other possible sources for the discrepancy observed. Nevertheless, some general remarks can be made on the surface-active behavior and surface adsorption of the Pluronic family.

1. At complete surface coverage, the area occupied by a copolymer molecule decreases with increasing temperature, meaning that the molecular conformation becomes more compact upon raising the temperature.
2. The area per copolymer molecule, A, increases with the number of EO units in the copolymer chain, showing a scaling relation $A \propto N_{EO}^{\nu}$ with $\nu \simeq 0.43$ (Fig. 10) [67]. A similar scaling relationship with $\nu \simeq 0.5$ was obtained for the simple nonionic surfactants—POE octylphenyl ethers, $C_8\Phi(EO)_i$ (i = 2–10)—when adsorbed at the air/water interface [155]. The plot in the latter case is also shown in Fig. 10.
3. An increase in the POP block length at the same POE block size gives rise to a decrease in the area occupied by each copolymer molecule, implying a more compact adsorbed layer for a larger POP block size.

4. The cmc/C_{20} ratio describes the competition between adsorption at the interface and micellization in the bulk phase. An increase in the cmc/C_{20} ratio indicates that micellization is inhibited more than adsorption, or that adsorption is facilitated more than micellization [156]. For conventional nonionic surfactants, this ratio has a value of 10–20 and increases with increasing POE length but decreases with increasing alkyl chain length and increasing temperature. In general, the Pluronic surfactants show similar but larger tendencies in so far as their cmc/C_{20} is concerned, as indicated by the remarkable variation of the cmc/C_{20} value shown in Table 9.
5. For Pluronic surfactants with the two breaks occurring on the σ-log C curve, the separation between them decreases with increasing temperature and finally vanishes at a high temperature of about 40 °C. Presumably, this is associated with the greatly enhanced micellization tendency with increasing temperature, whereas the corresponding increase in surface adsorption is more or less limited.

As previously noted, surface tension measurements are sensitive to the existence of surface-active impurities, showing typically a minimum in the vicinity of the cmc on the plot of surface tension against logarithmic concentration. Such a profound minimum in surface tension was reported for the Pluronic L-64 and F-127 systems [70,80]. After the removal of the "impurities" either by extraction with hexane or by fractionation, the surface tension minimum no longer exists. In general, the purified and unpurified Pluronic copolymers show similar micellar properties in solution.

H. Solubilization Behavior

Solubilization is directly related to micelle formation and can be defined as the enhanced solubility of a substance (called solubilizate) that is normally insoluble in the solvent, by reversible interaction with the micelles of a surfactant (called solubilizer) to form a thermodynamically stable isotropic solution. Solubilization is of great practical importance in many areas, such as detergency, emulsion polymerization, micellar catalysis, separation processes, and formulations involving water-insoluble ingredients. If the solubility of a dye that is normally water-insoluble is plotted against the surfactant concentration, the solubility is very slight before the cmc is reached, whereas above the cmc it increases approximately linearly with the surfactant concentration. Thus, dye solubilization constitutes a simple and convenient method for the determination of the cmc.

The solubilization measurements of *n*-hexane in Pluronic F-68 solution [130] indicate that at 25 °C, when the surface-active copolymer was molecularly dissolved, the solution appeared cloudy in the presence of hexane, because a macroemulsion (i.e., a dispersion of unsolubilized *n*-hexane droplets) was observed. At 35 °C, when the copolymer formed micelles, the solubilization of *n*-hexane did occur until a saturation limit (8.43 moles of hexane per mole Pluronic F-68) was reached. After that the transparent system turned into a cloudy macroemulsion. In addition, at 35 °C the hydrodynamic radius of the Pluronic F-68 micelle increased with the increase in the *n*-hexane uptake. The variation of the hydrodynamic radius of Pluronic L-64 micelles with the amount of solubilized toluene at different temperatures (25–43 °C) has been reported [169].

Pluronic surfactants show a unique feature that aromatic hydrocarbons are solubilized selectively over aliphatic hydrocarbons [122]. While only a 4-fold difference between the solubilized amounts of benzene and hexane has been observed for conventional surfactant micelles in aqueous solution, a 10–40-fold difference is obtained by using Pluronic copolymers, suggesting their potential application as an effective means for the separation of mixtures consisting of aromatic and aliphatic hydrocarbons. As a general rule, the solubilizates that are more compatible with the block constituting the micellar core, are solubilized in large quantities, and the substances that are poor solvents for that block show negligible solubilization.

Under certain conditions, water and *o*-xylene are selective solvents for POE and POP, respectively. The phase diagram of the ternary system, water–L-64–*o*-xylene, indicates that both a stable xylene in water and water in xylene one-phase region exist; these are located at the water-rich and oil-rich corners, respectively [104]. At the water corner, when the xylene content expressed as the xylene/Pluronic L-64 weight ratio is increased beyond a certain value, the solution becomes turbid or milky. Therefore, only a small region of the water corner is accessible to conventional light scattering measurements. Adding *o*-xylene (which is soluble in the POP core of the micelle) results in an increase in both the core size and the overall size, and a growth in the association number as well [104]. Therefore, the presence of a solubilizate increases the association tendency of the copolymer surfactant.

The solubilization in aqueous surfactant solutions can also be described quantitatively using the micelle-water partition coefficient (i.e., the concentration ratio at equilibrium of the solubilizate in the micelle and in the water). For Pluronic surfactants, the influence of the copolymer composition and structure on the solubilization was studied using the polycyclic aromatic hydrocarbons (naphthalene, phenanthrene, and pyrene) as solubilizates [63]. The linear EPE triblock copolymers are more effective in solubilization than branched tetronic copolymers, in agreement with the fact that the micellar core environment is less apolar for the branched copolymer structures. In all cases, the solubilized amount increases, as expected, with increasing POP content and increasing total molecular weight of the Pluronic copolymer, indicating that the POP portion is responsible for the solubilization process.

V. PHASE BEHAVIOR OF EO/PO POLYMERIC SURFACTANTS IN WATER

A. Introductory Remarks

Low molecular weight nonionic surfactants associate into micelles above the cmc in water and form different liquid crystal phases—cubic, hexagonal, and lamellar mesophases—at higher concentrations. In addition, the aqueous solution of POE nonionic surfactants shows a lower consolute temperature called the cloud point where the one-phase solution is separated into a dilute surfactant solution and a concentrated solution containing micelles. The phase behavior of nonionic hydrocarbon surfactants, particularly of the POE alkyl ether type $C_m(EO)_n$, has been extensively investigated [157]. In general, the phase equilibria examined show some common features independent of the detailed chemical structure of the polar groups. The phase equilibria are governed not only by the balance between the energetically favorable hydration of the polar groups and the energetically unfavorable contact between hydrocarbon and water, but also by the solubility behavior of the POE portion in water as a function of temperature which leads to the occurrence of a lower critical solution temperature. Pluronic polymeric surfactants can be viewed as analogs of conventional nonionic surfactants for which the short hydrocarbon chain is replaced by a POP block and the two ends carry the same polar POE groups. Therefore, in principle, similar phase properties could be expected for Pluronic surfactants, namely, 1) the selfassembly of the copolymer chains to form associated structures of different shapes (spheres, rods, etc.), 2) the growth and interactions of micelles to form order structures (e.g., liquid crystals), and 3) the occurrence of the cloud point for the solution phases and the melting point for the liquid crystalline phases. Note that the use of pure compounds is important in the phase diagram studies, because the type of distribution of the head group sizes (i.e., the polydispersity nature of EO groups present in commercial surfactants may have a significant effect on phase structures) [158]. For simple nonionic surfactants like $C_m(EO)_n$, the values of m and n can be well-defined and altered in their synthesis. Therefore, a systematic change of the surfactant chemical structure can be made in order to assess the influence of some important factors like the alkyl chain conformation and head group area on the structure and stability of the mesophases formed. Apparently, this kind of task for Pluronic copolymer surfactants would meet more difficulty and need more effort because of their composition polydispersity.

B. Phase Diagrams

The phase properties of the Pluronic/water systems over broad concentration and temperature ranges have started to attract considerable attention in the past

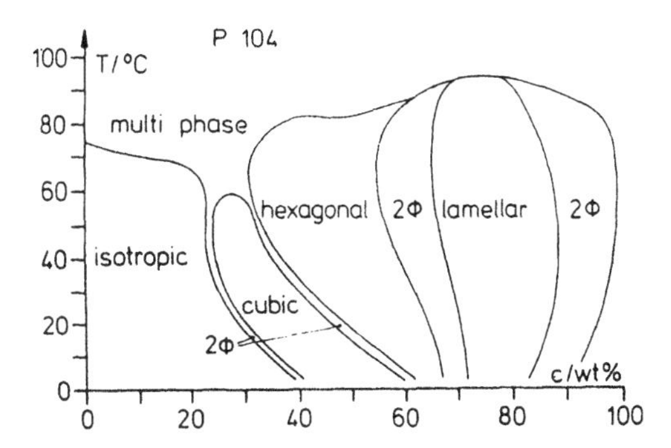

FIG. 11 Phase diagram of the system P-104/water. (*Source*: from [13], with permission.)

few years. Based on the SANS results, the Bayreuth group [12] first proposed the concept that a three-dimensional order (i.e., the existence of a cubic liquid crystalline phase) seems to be responsible for the gel formation of some Pluronic copolymers in aqueous solution. Later on, the Riso laboratory in Denmark and the Uppsala group in Sweden used SANS under shear to provide evidence that for concentrated copolymer solutions at elevated temperatures the Pluronic micelles crystallize in a body-centered cubic lattice [34,35]. Very recently, Hoffmann and coworkers [13] reported an extensive study on the phase diagrams of 12 Pluronic surfactants. In the meantime, the Lund research group completed a phase behavior study on seven members of the Pluronic family [60]. More phase diagrams of Pluronic surfactants with detailed information are expected to appear in the near future.

A typical phase diagram obtained for Pluronic P-104 in water [13] is given in Fig. 11, showing different kinds of mesophases. Below about 60 °C and with increasing copolymer concentration, the solution and liquid crystal phases appear in the following sequence:

spherical micelles → cubic phase → hexagonal phase → lamellar phase

In addition, a two-phase region exists between the two adjacent phases in the above sequence of phases. All the liquid crystalline phases melt below about 95 °C but the cubic phase at a lower temperature of about 60 °C. At temperatures higher than 95 °C, a multiphase region is observed where the separation into isotropic phases occurs. Figure 12 shows the phase diagram of Pluronic P-94 in water [60], displaying essentially similar features of the phase equilibria as Pluronic P-104.

Figure 13 shows the phase diagram of L-64 [60] which has the same EO/PO molar ratio as that of Pluronic P-94 and P-104, but a smaller total molecular

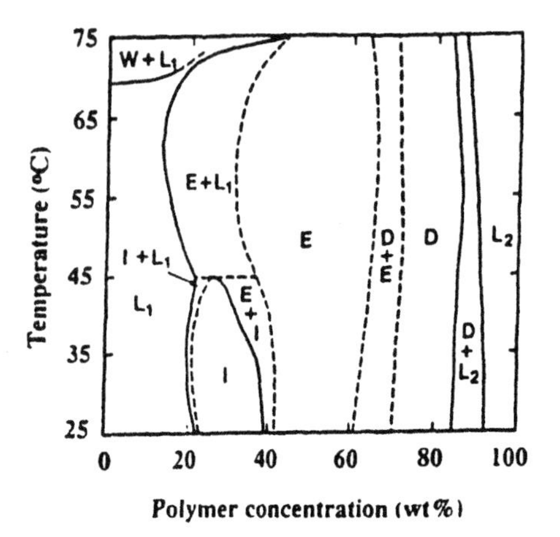

FIG. 12 Binary phase diagram of P-94–water system. L_1 represents aqueous polymer solution; W, very dilute polymer solution; L_2, concentrated polymer solution; E, hexagonal liquid crystalline phase; I and V, cubic liquid crystalline phases; D, lamellar liquid crystalline phase. (*Source*: from [60], with permission.)

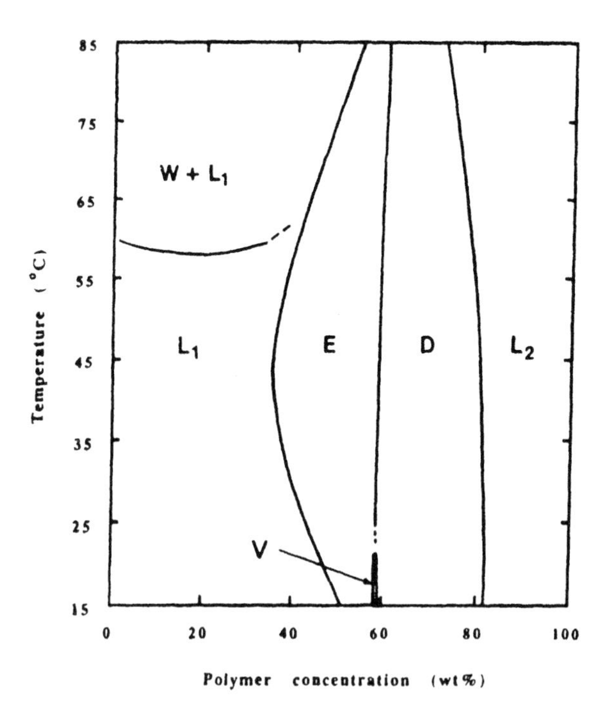

FIG. 13 Binary phase diagram of L-64–water system. Symbols are the same as in Figure 12. The two-phase regions, except the liquid-liquid one ($W + L_1$), are narrow and are omitted. (*Source*: from [60], with permission.)

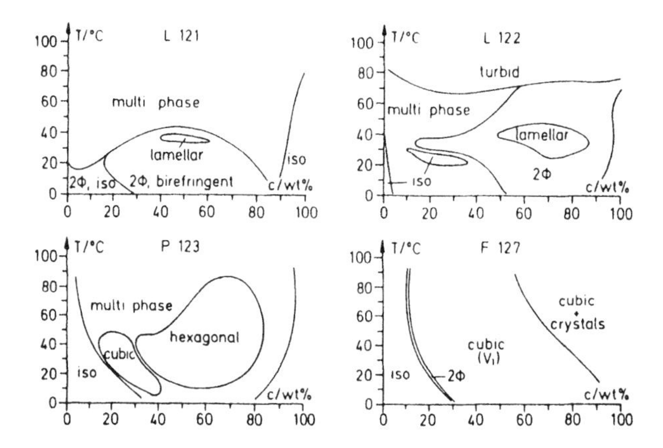

FIG. 14 Phase diagrams of the systems of $(EO)_x(PO)_y(EO)_x$ block copolymers with $y = 69$ and x increasing from 5 to 106. (*Source*: from [13], with permission.)

weight of 2900. Two remarkable differences in the phase behavior are observed for this system. First, the hexagonal phase, instead of the cubic phase, appeared as the first liquid crystalline mesophase. Second, a cubic phase (V) was detected between the hexagonal and the lamellar mesophases within very restricted ranges of concentration and temperature (Fig. 13). In addition, the cubic phase consists of a bicontinuous network with water and polymeric surfactant forming the bicontinuous zones. Thus, this cubic phase differs from the normal cubic phase obtained for Pluronic P-94 or P-104 in water, which consists of small micelles and exists between the isotropic micellar solution and the hexagonal phase. The authors [60] stressed that the range of its existence is so narrow that the preparation of single cubic phase (V) of Pluronic L-64 is not easy. Interestingly, when examining the phase behavior of PE 6400 (virtually identical with Pluronic L-64 in composition) in water, a similar phase diagram was obtained [13], showing the appearance of a hexagonal phase as the first disorder-to-order transition, but no cubic mesophase (V) was found. The difficulty in its preparation, as mentioned above, may account for this disagreement. Thus, even though the EO/PO molar ratio is the same, the variation in the total chain length of the copolymer molecule, as indicated by these results, has an important influence on the type and structure of the mesophases involved.

To evaluate the effect of the POE block length on phase properties, Fig. 14 gives the phase diagrams of those Pluronic surfactants ($E_mP_nE_m$) which have the same POP block size ($n = 69$) but a varying m value from 5 to 106. With the same purpose of comparison, Figs. 15 and 16 are shown for the Pluronic surfactants having constant but smaller POP block lengths (i.e., with $n = 30$

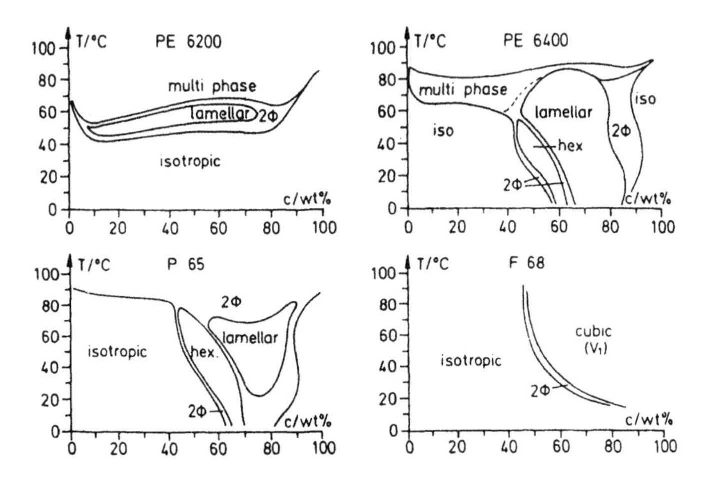

FIG. 15 Phase diagrams of the systems of $(EO)_x(PO)_y(EO)_x$ block copolymers with $y = 30$ and x increasing from 5 to 80. (*Source*: from [13], with permission.)

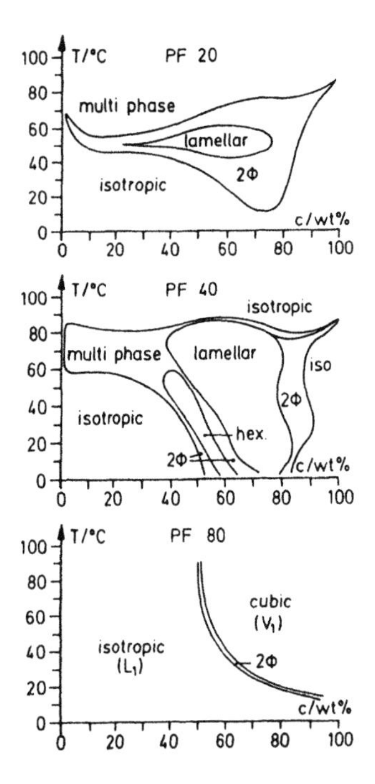

FIG. 16 Phase diagrams of the systems of $(EO)_x(PO)_y(EO)_x$ block copolymers with $y = 27$ and x increasing from 5 to 73. (*Source*: from [13], with permission.)

and m = 5–80, and n = 27 and m = 5–73, respectively). When examining Pluronic L-31, L-61, and L-62 in water over broad concentration and temperature ranges, no liquid crystalline phase was detected at all [60]. On the basis of these phase behavior studies several general remarks can be made:

1. It seems that for Pluronic surfactants a threshold value of total molecular weight (about 2000) exists, below which no liquid crystalline phase forms up to the cloud point [60], as indicated by the phase properties of Pluronic L-31, L-61, L-62 [60], and PF-20 [13] in water.
2. The composition of the Pluronic copolymer (i.e., the EO/PO molar ratio and the total molecular weight) is expected to have a large influence on its phase behavior. Based on their phase diagram studies, the Bayreuth group [13] concluded that the sequence of the mesophases observed is determined mainly by the EO/PO molar ratio of the copolymer molecule (i.e., the m/n value). The greater the m/n value, the more is the possible number of mesophases which could exist. To be specific, when $m/n \geq 0.5$, spherical micelles are first formed when $C >$ cmc. A disorder-to-order transition occurs at higher concentrations, leading to the formation of a cubic phase. With further increase in concentration, transitions to hexagonal and to lamellar mesophases appear in sequence. When m/n is reduced to about 0.25, the hexagonal phase becomes the first liquid crystalline mesophase, whereas the lamellar phase would appear as the first ordered phase with $m/n \simeq 0.15$.

   Similarly, the dependence of the mesophase sequence on the number of EO groups was reported for nonionic hydrocarbon surfactants [158]. For example, for $C_{16}(EO)_m$ in water, cubic phases were found only when $m >$ 8, while extensive lamellar phase regions occurred only with small EO groups (i.e., $m < 5$).
3. For conventional nonionic surfactants, the intra- and intermicellar interactions are thought to be responsible for the micellar shape and the mesophase structure, respectively [158]. Accordingly, the possible phase sequence in the disordered and ordered solutions is determined by the curvature of the micellar surface and the surfactant volume fraction (see Fig. 2 in [158]). In the case of Pluronic polymeric surfactants, the cloud point is an important parameter which can be considered to be approximately parallel to the curvature of the micelle formed, as both of them are related to the EO/PO ratio and the total molar mass. In this connection, Zhang [60] correlated the mesophase structure with the cloud point of the polymeric surfactant instead of the micellar curvature and gave a schematic description (Fig. 17).
4. One characteristic feature of the phase diagrams (Figs. 14–16) is the diagonal track of the phase boundaries. Thus, the mesophases of Pluronic

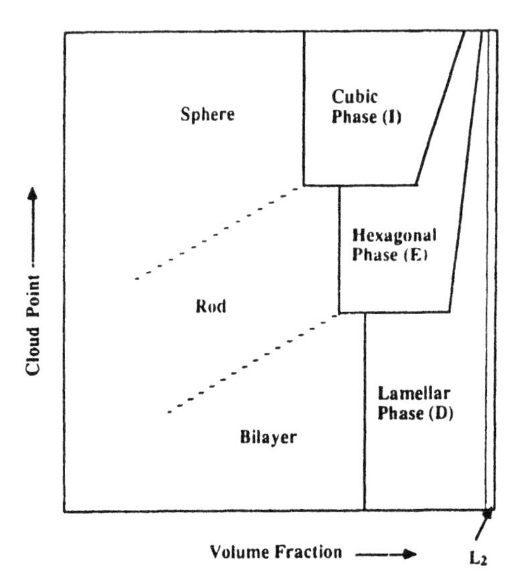

FIG. 17 Schematic illustration of mesophase structures versus volume fraction and cloud point of the polymer. (*Source*: from [60], with permission.)

surfactants show thermoreversible transitions (e.g., from the isotropic solution to the ordered phase or from one mesophase to the other) as a function of temperature at a constant concentration. Such a thermotropic behavior is of importance, and is closely related to the thermally reversible gel formation of the Pluronic systems.

Experimentally, liquid crystalline phases are usually identified by the textures under the polarization microscope. In general, the textures of block copolymers are not as distinct as for simple surfactants, and the interior of copolymer micelles are much less ordered than that of simple hydrocarbon surfactants [13,60]. Moreover, the large-scale domain structure is less well developed for the copolymers. Consequently, tempering the sample for a longer time is often necessary to obtain a distinct texture.

C. Small-Angle Neutron Scattering Studies on Phase Behavior and Phase Structure

When dissolved in deuterated water, the Pluronic copolymer forms a homogeneous solution with good scattering contrast between the protonated part and the deuterated part of the system, while the contrast between POP and POE is negligible. Thus, the small angle neutron scattering technique is very useful in characterizing both the micelle and the mesophase structures of Pluronic surfactants. The Bayreuth group first reported their SANS results on Pluronic F-127 and P-123 at different temperatures, both in the solution and the gel states [12]. The strong correlation peak occurring in the scattering patterns at higher concentrations is an indication of a nearest-neighbor order in the two systems studied. Furthermore, the scattering profile of Pluronic F-127 in the gel state reveals the presence of a long-range order, and thereby the existence of, probably, a cubic liquid crystalline phase. Later, extensive SANS studies were carried out, mainly on Pluronic F-88 and P-85, by the Riso laboratory, focusing on the structural properties of the micellar phase and the liquid crystalline phase as a function of temperature [34–41].

In the low temperature region (i.e., much below ambient temperatures), the neutron scattering profiles show comparatively weak q dependence with q being the magnitude of the scattering vector, and a small absolute scattered intensity. Such scattering features are consistent with fully dissolved Gaussian copolymer molecules [36]. Thus, the radius of gyration was found to be 1.7 and 2.3 nm for Pluronic P-85 and F-88, respectively, in agreement with the corresponding hydrodynamic radius values of 1.8 and 2.9 nm obtained from dynamic light scattering measurements [17,18].

As noted in Section IV, the POP part of the copolymer chain becomes virtually insoluble in water as the temperature is close to ambient, leading to the formation of spherical micelles. The micelles are in the form of a core dominated by the POP block and surrounded by a protective outer layer of highly hydrated POE blocks. Within a certain temperature range, the micelles are in equilibrium with the unimers and the equilibrium shifts in favor of micelles with increasing temperature. Consequently, a marked increase in scattered intensity of neutrons appears when the temperature is well above the cmt. In essence, the measured SANS profile is dominated by the large micelles, since the scattering ability of a single particle increases with its volume squared.

As the copolymer concentration or the temperature is further increased, the interactions between neighboring micelles become increasingly more important, thus yielding an increasingly pronounced correlation peak in the scattering function. In the meantime, the neutron scattering profile in the high q region remains unchanged. Accordingly, under such circumstances the number of micelles is increased but individual micelles are little affected.

The scattering function, $I(q)$, can be expressed in terms of the product of the form factor, $P(q)$, and the structure factor, $S(q)$, which reflect the intramicellar and intermicellar interferences respectively. Thus, we have

$$I(q) = nKS(q)P(q) \tag{10}$$

where n is the number density of micelles and K is the contrast factor. Based on the hard-sphere model, Mortensen et al. [36,37] established an approach to

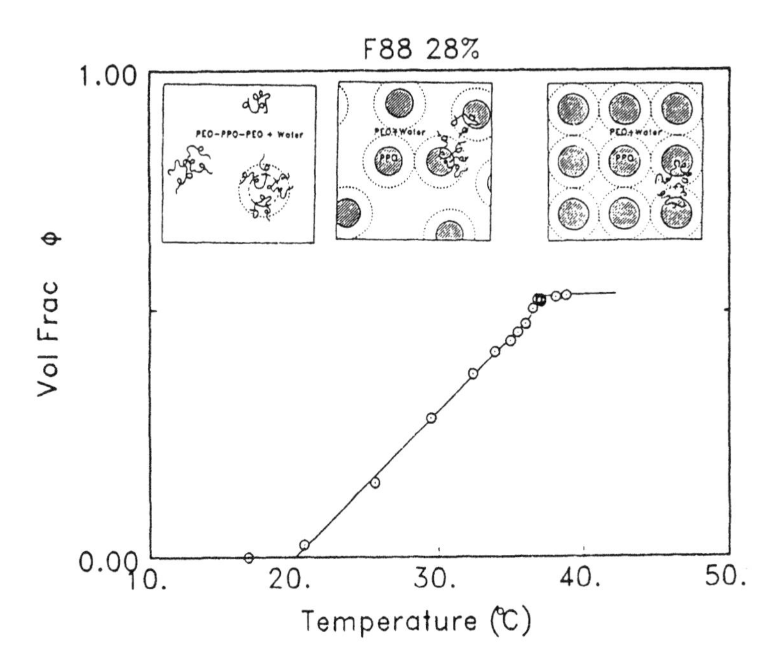

FIG. 18 Micellar volume fraction, as obtained by Percus-Yevick fits to the scattering data of 28% F-88. The inset shows the structure of F-88 in the three temperature ranges: low-temperature unimers in solution, intermediate-temperature micellar liquid, and high-temperature cubic colloidal crystal. (*Source*: from [38], with permission.)

determine the three characteristic quantities: the core radius of the micelle, r_c; the hard-sphere interaction radius, R_{HS}; and the hard-sphere volume fraction, Φ. By assuming a monodisperse hard-sphere model with a dominating contribution of the dense spherical core to the scattering for the form factor and by using the Percus-Yevick approximation [159] to treat the structure factor, the three parameters can be evaluated by the model to fit the experimentally measured scattering functions. The r_c and R_{HS} values thus obtained are essentially independent of copolymer concentration, but increase with temperature as required by the temperature-induced increase in association number. For example, for Pluronic P-85 in water, r_c and R_{HS} are 3.8 and 6.0 nm at 20 °C, and increase to 5.1 and 7.5 nm at 50 °C, respectively. Importantly, the micellar volume fraction obtained by the model fit increases linearly with both temperature and concentration. For example, Fig. 18 shows the results obtained by fitting the hard sphere Percus-Yevick model to the experimental neutron scattering data of a 28% Pluronic F-88 solution [38]. The volume fraction of micelles increases linearly with temperature until a plateau value $\Phi = 0.53$ is reached where the micellar liquid undergoes a first-order phase transition to a cubic crystal. Such a "inverse melting transition" showing a sol-to-gel transition in the external appearance is, in essence, hard sphere crystallization [160]. The inset in Fig. 18 shows the three characteristic temperature ranges of the 28% Pluronic F-88 solution and the corresponding structures of the disordered solution phase and the ordered cubic phase. At temperatures below 20 °C, the copolymer exists in the form of single Gaussian coils and only a few micelles may appear. At intermediate temperatures, the dehydration of the POP block causes micelle formation on a noticeable scale. Upon raising the temperature in this region, both the number and the size of micelles increase and therefore the micellar volume fraction also increases. When the temperature is further increased to reach the critical volume fraction ($\Phi = 0.53$), the cubic phase appears as a result of the close packing of hard-sphere micelles. This inflection temperature corresponds to the sol-gel transition temperature in the Pluronic system (Fig. 18).

Experimentally, for a Pluronic solution in the rest state, only a little change in the neutron scattering patterns has been detected at or above the inflection temperature (i.e., the crystallization temperature). However, when a small shear field is applied, the azimuthally isotropic scattering pattern abruptly transforms into a hexagonal pattern of Bragg reflections [34,35]. In other words, upon exposure to shear the polycrystalline material aligns effectively and transforms into a single crystal. For a 25% Pluronic P-85 solution at 27 °C (Fig. 19b), the occurrence of a six-fold symmetric Bragg peaks corresponds to the plane of the body-centered cubic (bcc) crystal with the (110)-type reflections (111) plane. Thus, the bcc nature of the colloid crystal phase has been proven for both the Pluronic P-85 and F-88 samples under shear [34,35]. To demonstrate the temperature-induced effect, the neutron scattering pattern for Pluronic P-85 at 25 °C as well as the schematic description of the disordered liquid phase of spherical micelles is shown in Fig. 19.

A sphere-to-rod transition in the micellar shape of Pluronic P-85 takes place around 70 °C. Correspondingly, the neutron scattering characteristics of Pluronic P-85 in D_2O change significantly. At low concentrations, the correlation peak is no longer seen. At high concentrations the Bragg reflections disappear, thus indicating the melting of cubic crystals. Moreover, above 70 °C, the scattering function in the low concentration region indicates a structural change for the Pluronic P-85 micelles from spheres to larger prolate ellipsoids. As previously noted, the shape transition occurs when the micellar core diameter becomes comparable to the fully extended length of the POP block. For concentrated Pluronic P-85 solutions, a further increase in temperature leads to the appearance of another crystalline mesophase consisting of hexagonally ordered rodlike micelles, as shown by the corresponding neutron scattering pattern obtained under shear (Fig. 19, bottom right). Interestingly, the two crystalline phases described are separated by a liquid phase of prolate or rodlike micelles. On the lower temper-

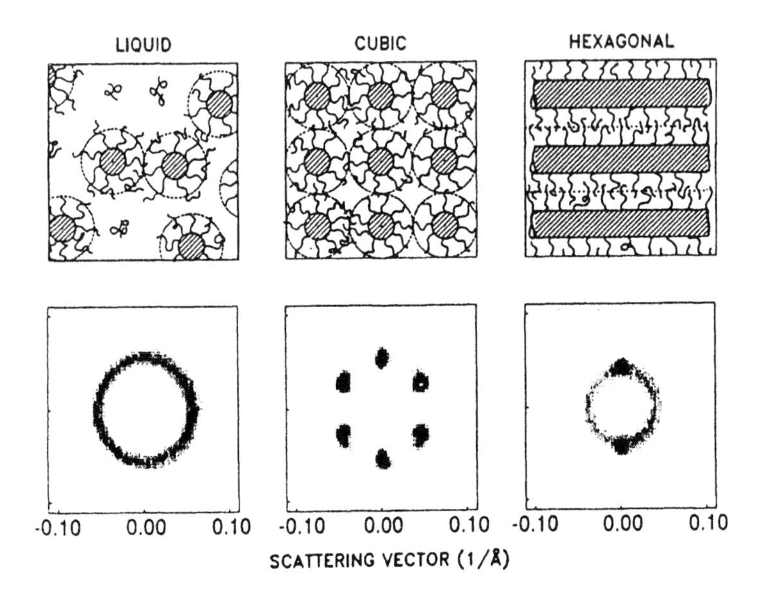

FIG. 19 Top row: Schematic illustration of interacting micelles, spherical at low temperature, characterized by the core-radius R_c and hard-sphere interacting radius R_{hs}, and rod-like at high temperature. Bottom row: Two-dimensional scattering functions as obtained perpendicular to the shear plane of 25% solution $E_{25}P_{40}E_{25}$. The three columns represent P-85 at T = 25 °C, T = 27 °C, and T = 68 °C. (*Source*: from [39], with permission.)

ature side, the bcc crystalline phase is seen, and on the other side, a hexagonal mesophase is detected. At higher temperatures of about 80 °C, the latter melts and a new liquid phase appears. The phase behavior and the phase structure at varying temperatures and concentrations for Pluronic P-85 in water, as derived from SANS experiments, are summarized in Fig. 20 [35,36]. In addition, pressure improves the solvent quality of water, and therefore favors the decomposition of micelles and the melting of the micellar crystal [40]. In general, as the influence of pressure on the micellar system is concerned, increasing pressure is equivalent to decreasing temperature.

D. Chain Architecture and Phase Behavior

The chain architecture of block copolymers has a large influence on their association characteristics in selective solvents. A symmetrical triblock copolymer, when dissolved in a solvent selective toward the central block, shows a variety of possible associated structures. Also, in this case a complex phase behavior could be expected, depending, to a large extent, on the block length ratio and the solvent quality variation with temperature. To date, a large number

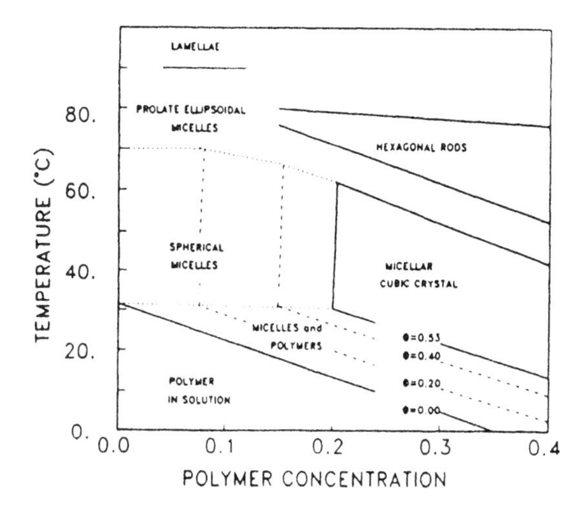

FIG. 20 Temperature-concentration phase diagram showing the characteristics of $E_{25}P_{40}E_{25}$ fully dissolved in D_2O. In the low-concentration, low-temperature range, the polymers are dissolved Gaussian chains. Above the $\Phi = 0$ line of critical micellation concentration/temperature, a liquid of micelles is formed with micellar volume fraction (Φ) as shown by contour lines in the figure. When the $\Phi = 0.53$ line is crossed, the micellar liquid "freezes" into a hard-sphere cubic crystal. (*Source*: from [36], with permission.)

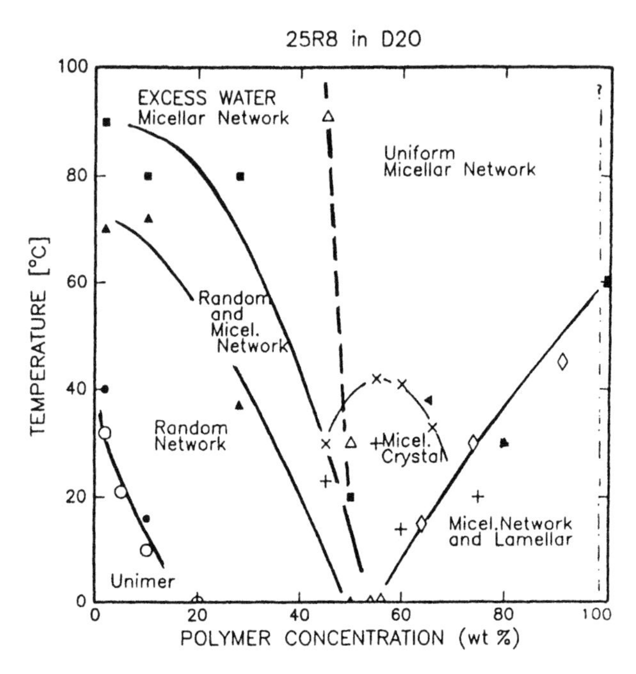

FIG. 21 Phase diagram of the Pluronic R system 25R8. Closed symbols refer to neutron scattering results, open symbols to light scattering, and crosses to rheological data. (*Source*: from [42], with permission.)

of phase diagrams are obtained for the EPE type block copolymers in aqueous medium, but few phase behavior studies are available for the inverse type of triblock copolymers [42,109]. Figure 21 shows the phase diagram of Pluronic 25R8 ($P_{15}E_{156}P_{15}$) in water, obtained by SANS, static and dynamic light scattering, and rheological measurements [42]. This triblock copolymer consists of a long hydrophilic central block and two short hydrophobic end blocks. Several interesting but complicated features are seen from the phase diagram obtained:

1. At low temperatures and in dilute region, the copolymer exists in the form of unimers with a Gaussian conformation as evidenced by the scattering experiments.
2. Upon increasing the concentration or raising the temperature, the intermolecular association of the short POP blocks yields small clusters which are interconnected by the long POE block to form branched structures. The authors [42] called them random network (Fig. 22) and the branched structures may exist within a certain range of temperature and concentration (Fig. 21). They are not permanent structures and may differ greatly in size.

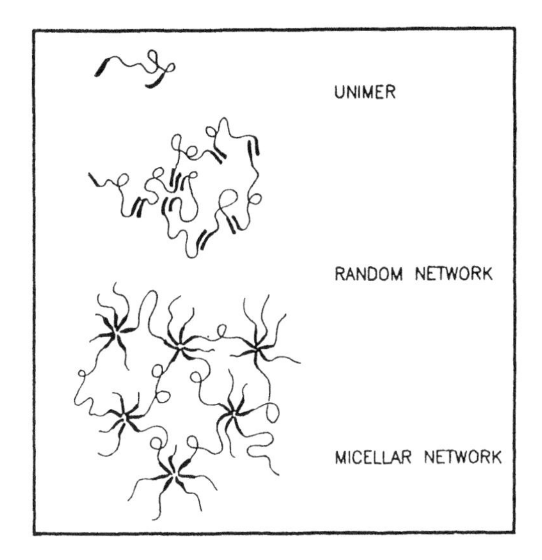

FIG. 22 Schematic describing the different phases, random coil, random network, and micellar network. (*Source*: from [42], with permission.)

3. At higher concentrations or higher temperatures, spherical micelles containing a more or less well-defined POP core are formed, as suggested by the neutron scattering results. The long hydrophilic POE portion functions as a bridge to bring the micelles together to form the isotropic micellar network (Fig. 22). For concentrations above 50 wt%, this kind of randomly oriented, large isotropic domain may constitute the whole sample, making it liquid-like in appearance (Fig. 21).
4. Within a restricted area in the T-C diagram (e.g., at concentrations between 50 and 70 wt%, and below 40 °C), a micellar mesophase, presumably of the cubic nature, is formed as a result of hard sphere crystallization, showing a solid-like elastic gel appearance.
5. At the highest copolymer concentrations and low temperatures, a two-phase system with elastic properties is formed due to the coexistence of a swollen lamellar domain with the micellar network.

The PEP-type polymeric surfactants exhibit a much lower cloud point in water than the EPE counterparts. For example, a 10 wt% aqueous Pluronic 25R8 solution shows a cloud temperature of 16 °C [161], although it contains 80% by weight of POE. At C = 45 wt%, the Pluronic 25R8 solution forms a milky liquid below 55 °C [42]; consequently, only limited T-C areas are accessible to light scattering measurements (indicated by the open symbols in Fig. 21).

Polyoxyethylene dialkyl ethers are similar in the structure to the PEP triblock copolymers with the two terminal POP blocks being replaced by alkyl groups. With a long POE chain (10,000–30,000 mw) as the central part and an alkyl group containing 16–21 carbon atoms on each side, this type of compound forms a water-swollen gel with the micellar network structure instead of giving a homogeneous solution. Thus, this gel is substantially different from the Pluronic gels formed by close packing. The swollen POE chains are mainly responsible for the elasticity and the physical crosslinks are carried out by the association of the alkyl groups into nanodomains like micellar cores. The Manchester group [91], using NMR and DSC, found that the nanodomains of alkyl chains undergo a solid/liquid transition at temperatures similar to the melting points of normal aliphatic hydrocarbons with 3–4 carbon atoms less than the alkyl chains.

E. Phase Behavior in Nonaqueous Solvents

The published work on the POE/POP type block copolymers in solvents other than water is rather rare. Lindman and coworkers [44,45] studied the phase behavior of this class of block copolymers in a variety of polar solvents. They used two diblock copolymers with a EO/PO molar ratio of 0.8 and 0.33, and a molar mass of 2920 and 3440, respectively. They found that the copolymers are soluble in most solvents, but the clouding behavior is observed only in water and formamide. The cloud point is higher in formamide solution than in aqueous solution, which can be interpreted by the smaller difference in polarity between the copolymer and formamide as compared with that between the copolymer and water. In general, the phase behavior of the copolymer in formamide is qualitatively similar to the behavior in water, both with and without salt added.

VI. GELLING PROPERTIES OF EO/PO POLYMERIC SURFACTANTS IN WATER

A. Thermoreversible Gel Formation

In addition to the strongly temperature-dependent micellization behavior in the dilute regime, the second most distinct property of Pluronic systems is the heat-induced gel formation at higher concentrations. For many Pluronic copolymers at high concentrations in aqueous solution, a gel region exists within a certain temperature range. Such a thermal gelation phenomenon has been known for quite long time and is of potential importance particularly in pharmaceutical applications [1]. As an empirical rule, Pluronic surfactants in which the hydrophobic POP block has a molecular weight of 940–1100 and the POE portion constitutes from 15 to 80% of the total weight, do not form gels in aqueous solution [139,140]. Only those Pluronic copolymers in which the molar mass

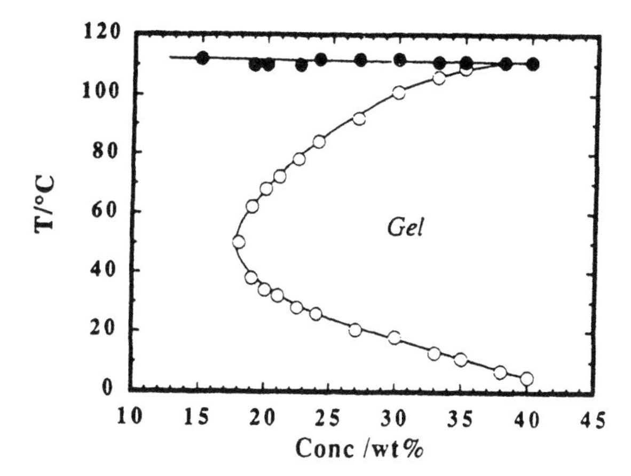

FIG. 23 Phase behavior of the $E_{99}P_{65}E_{99}$ /water system. The solid circles refer to the cloud point. (*Source*: from [52], with permission.)

of the hydrophobe is 1750 or more, are able to form gels at appropriate temperatures. In addition, as the molar mass of the POP block increases, the minimum copolymer concentration required for forming a gel decreases from 60% to about 20%. In general, the larger the POP block and the greater the POE content, the greater is the gelling ability of the Pluronic copolymer (i.e., less copolymer is needed to form a gel).

Figure 23 shows a typical gelling behavior of Pluronic F-127 ($E_{99}P_{65}E_{99}$) in water as a function of temperature and concentration [50]. The cloud points of aqueous Pluronic F-127 solutions at different concentrations are also shown. The gel region appears at concentrations above 18 wt% with a corresponding temperature interval (Fig. 23). The temperature range for the gel phase increases with increasing copolymer concentration. Figure 23 indicates that at a given solution concentration, a temperature increase causes the sol-to-gel transition to occur first (i.e., the system changes from a fluid to an immobile state). The system is characterized in the gel region by a high elastic modulus and a low loss modulus. When the temperature is further increased to meet the upper gel boundary, the system transforms back to a low-viscous solution. The transitions are thermoreversible, showing similar mechanical properties in both heating and cooling runs.

Although the gelation and the melting mechanisms at the lower and upper transition boundaries and the structure of the gels are not yet well understood, it is believed that the gel formation is generally related to a close-packed array of Pluronic micelles. As described previously, SANS studies [12,34–41] provide experimental evidence on Pluronic P-85 and F-88 that the gel state is a crystalline phase of bcc-ordered spherical micelles. Thus, the mechanical properties in the gel state are connected with the existing liquid crystalline mesophases. In this connection, Hoffmann and coworkers [13] pointed out that similar to conventional nonionic surfactants, the cubic and hexagonal mesophases of Pluronic copolymers exhibit a yield stress that is high enough to prevent the flow of the system under the gravitational field. But, the lamellar mesophases show no yield stress or a much smaller one. At ambient temperatures, for example, the 55-wt% Pluronic P-104 solution that forms a hexagonal phase, does not flow over several months, while the 70-wt% solution gives a lamellar phase and flows within 1–2 weeks. However, with increasing temperature two sol-to-gel transitions were observed on Pluronic P-94 in water [24]. While the low temperature gel with a minimum copolymer concentration of 24.5 wt% reveals a large elasticity, the high temperature gel formed at concentrations as low as 1–2 wt% shows only a small elastic modulus. It is likely that different mechanisms are involved for these two transitions. The authors [24] further suggested that the gelation seems to be dominated not by crystallization but by packing and overlapping between micelles.

B. Phase Diagrams in Terms of Unimer, Micelle, and Gel Regions

Phase diagrams are presented in the preceding section with the attention being focused on the phase structure (i.e., the nature of the liquid crystalline mesophase involved). Now, phase diagrams for the Pluronic/water systems are given to show the combined concentration and temperature conditions under which the unimer, micelle, and gel regions are found. In other words, more attention is paid to the outer appearance of the system rather than the internal structure. To date, this kind of informative phase diagrams of Pluronic copolymers concentrating on the gelling properties is limited in number and has become available only recently.

1. Phase Diagram of Pluronic P-85 in Water

Figure 24 shows the phase diagram for Pluronic P-85 in water, obtained by combining a variety of experimental methods, including SAXS, ultrasonic speed measurements, DSC, low shear viscosimetry, and light transmission measurements [23]. The Pluronic P-85 sample used was purified by the procedure described in [70]. In Fig. 24, there is a broad transition region as marked by vertical hatching where unimers and micelles coexist. The unimer region is located below the transition regime. The SAXS results of a 1 wt% copolymer solution indicates the spherical star-shaped structure of the Pluronic P-85 micelles consisting of a compact POP core with POE arms in the outer layer. At concentrations below 24 wt% and above the transition range, the micelles are

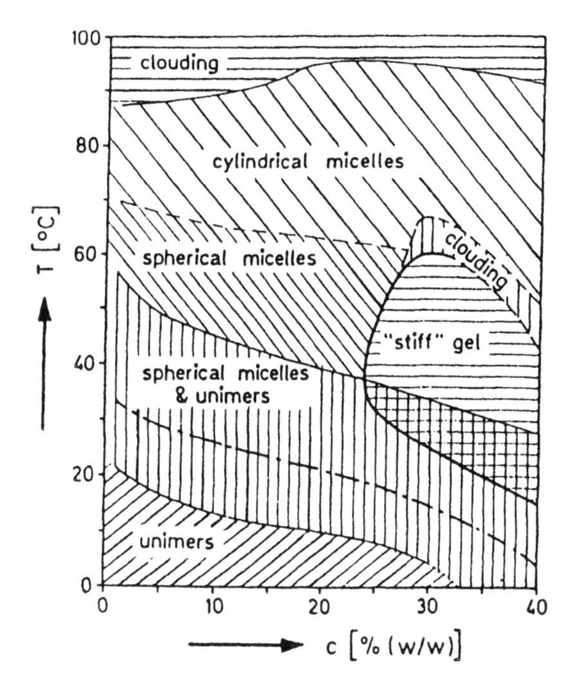

FIG. 24 Phase diagram for aqueous P-85 solutions. (*Source*: from [23], with permission.)

found to grow up to about 70 °C. Cylindrical micelles with $L/d \simeq 7$ are formed at dilute concentrations ($C < 1\%$) in the temperature range 70–85 °C. A further increase in temperature leads to the clouding of the system.

The SAXS results at different concentrations (25–35 wt%) and at 40 °C suggest that with increasing concentration, the micelles become increasingly deviated from the spherical shape [23]. Thus, the SAXS results seem to be in contradiction with the hard sphere interaction model which is used in the analysis of the SANS data. For the Pluronic P-85/H_2O or P-85/D_2O system, the scattering contrast is considerably different between SAXS and SANS; it is therefore understandable that some kind of variation may occur in the final results given by the two methods, but there should be no substantial disagreement.

A stiff gel region exists at concentrations above 24 wt% (Fig. 24). Importantly, the phase boundaries of this region established by low shear viscosimetry are in good agreement with those determined by the other techniques which do not apply any shear field. Above the stiff gel region there is a small two-phase region showing a clouding appearance. At the bottom, the stiff gel phase overlaps with the upper part of the transition region. The SAXS results in the gel region without shear applied, give no evidence for a micellar cubic crystal, but suggest the likely contribution from a lamellar structure at high concentrations [23]. As the authors noted, further structural studies are required in both the gel phase and the cylindrical micelle region. Nevertheless, when comparing Figs. 20 and 24 for the Pluronic P-85/water system as obtained by different groups using different techniques, one would find that they are virtually similar to each other in the main aspects, thus demonstrating some connection existing between gelation and micellar crystalline phases.

2. Phase Diagram of Pluronic P-94 in Water

The Pluronic P-94 sample contains about 6% of a lower molecular weight portion with higher PO content. After purification, the copolymer composition, as determined by ^{1}H and ^{13}C NMR, was $E_{28}P_{48}E_{28}$ [24] instead of the nominal composition $E_{21}P_{47}E_{21}$. As noted before, light scattering, surface tension, and cloud point results are sensitive to the impurities. But, unpurified and purified Pluronic P-94 samples generally exhibit identical behaviors in viscoelasticity and DSC measurements. Figure 25 shows the phase diagram for a purified Pluronic P-94 sample at concentrations up to 40% [24], as determined by a combination of viscoelasticity, differential scanning calorimetry, and cloud point techniques. Similar to Pluronic P-85 in water, aqueous Pluronic P-94 solutions show a broad transition temperature region where micelles coexist with unimers. Rheological measurements indicate that aqueous Pluronic P-94 solutions un-

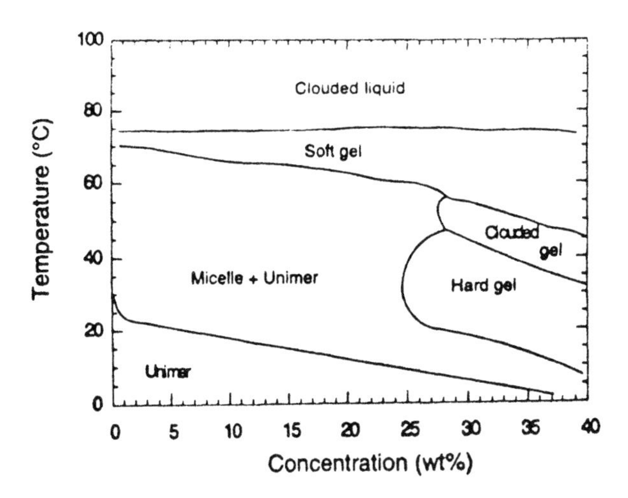

FIG. 25 Diagram showing the combined concentration and temperature conditions where P-94 is found as unimers, micelles, and gels in water, as determined by use of viscoelastic, scanning calorimetry, and cloud point techniques. (*Source*: from [24], with permission.)

dergo complex sol-to-gel transitions with increasing temperature. A gel state appears at a critical concentration of 24.5 wt% and a critical temperature of 32 °C (Fig. 25). Above this concentration, there is a hard gel region exhibiting a high elastic modulus of $(1\text{–}4) \times 10^4$ Pa. This hard gel region occurs at concentrations and temperatures similar to the cubic crystalline region of Pluronic P-85 as observed by SANS measurements (see Fig. 20) or to the stiff gel region for Pluronic P-85 observed by other techniques (see Fig. 24). At concentrations between 24.5 and 28 wt%, the hard gel melts at higher temperatures and a low modulus viscoelastic liquid is formed. A further increase in temperature results in the formation of a soft gel with a small modulus between 1–50 Pa. Finally, the soft gel region ends with a two-phase liquid state having a cloud temperature of about 75 °C that is independent of concentration. Moreover, at concentrations above 28 wt%, a cloudy gel region consisting of two phases and showing a decreased elastic modulus with temperature is observed. The cloudy gel region separates the hard and soft gel regions. Pluronic P-94 is similar to P-85 in the total molecular weight (= 4600 g/mol) but has a higher POP content. The soft gel region of Pluronic P-94 occurs at high temperatures similar to that for a hexagonal mesophase region of Pluronic P-85 (Fig. 20), but shows a large difference that only a low concentration of 1–2 wt% is needed for Pluronic P-94 to form the soft gel, while with Pluronic P-85 a minimum concentration of 14 wt% is required for the hexagonal crystalline region. Besides, the liquid channel for Pluronic P-85 to separate the cubic and hexagonal phases (Fig. 20) is replaced by the cloudy gel region of Pluronic P-94 (Fig. 25).

In brief, two sol-to-gel transitions are observed for Pluronic P-94 with increasing temperature and then followed by phase separation. Hvidt et al. [24] suggested that the hard gel consists of amorphous close-packed spheres, implying that the micelles may not necessarily form a cubic crystal, and that the large elastic shear modulus observed can be attributed to the increased surface area of the hydrophobic POP core against water under shear deformation. The soft gel is formed at high temperatures and shows a small shear modulus. The authors [24] speculated that the origin is possibly due to the hindered rotation of overlapping rodlike micelles rather than crystallization. On the other hand, comparing Figs. 12 and 25, both of which are obtained for Pluronic P-94 in water, shows that the hard and soft gel regions are more or less associated with the occurrence of the cubic and hexagonal crystalline mesophases, respectively. To have a deep insight into the complicated gelling properties, more precise studies on the phase behavior and phase structure for more Pluronic systems are required.

C. Effect of Additives on Gel Region

For an aqueous Pluronic solution whether a gel is formed or not, it generally depends on how close to the critical value (Φ = 0.53) the effective volume fraction of the micelles is. The latter quantity, in turn, is determined by the number and size of micelles. Therefore, any factors which influence either the micellization process or the micellar parameters, are expected to change the stability region of the gel.

1. Effect of Homopolymers

Malmsten and Lindman [52] studied the effect of homopolymers (i.e., POE and POP) on the gel formation of Pluronic F-127 in aqueous medium. POE reduces the gel region of Pluronic F-127 and the gel region vanishes completely at a sufficiently high POE concentration. The amount of POE required for "melting" the gel depends on the copolymer concentration and becomes smaller with increasing POE molecular weight. While POE of low molecular weight (400 and 600 mw) hardly affects the gelation behavior, the high molecular weight POE (200,000 mw) induces phase separation instead of gel melting. POE of an intermediate molecular weight (e.g., 6000 mw) increases the gelation temperature and decreases the melting temperature. In strong contrast, POP (400 mw) increases the stability region of the Pluronic F-127 gel. However, POP (4000 mw) is too large to be solubilized. For example, 1 wt% POP (4000 mw) cannot be solubilized by a 20 wt% F-127 solution.

2. Effect of Salts and Hydrocarbons

Inorganic salts like NaCl, KCl, and Na_2SO_4, decrease the sol-to-gel transition temperature of Pluronic copolymers, while urea and ethyl alcohol show the opposite effect [132]. The Lund group [50] further pointed out that the presence of a typical "salting-out" agent like NaCl shifts the whole gel region as well as the cloud temperature to lower temperatures, whereas adding a "salting-in" agent moves the gel region and the cloud point to higher temperatures.

Aromatic hydrocarbons are solubilized selectively over aliphatic hydrocarbons in aqueous Pluronic solutions [122]. When *p*-xylene is solubilized with a xylene/copolymer molar ratio of 10, the lower temperature gel boundary of Pluronic F-127 shifts to lower temperatures [50]. In the meantime, a small increase in the upper gel boundary and a small decrease in the cloud point are observed.

3. Effect of Ionic Surfactants

Recently a number of studies have been devoted to the interactions of Pluronic copolymers with ionic surfactants, in particular, with the anionic surfactant sodium dodecylsulfate (SDS) [14–16,26,61,132]. Almgren et al. [26] studied the interaction of Pluronic L-64 and F-68 with SDS in the dilute regime using ^{13}C NMR and fluorescence quenching techniques. They found that the Pluronic copolymers form mixed micelles with SDS at concentrations well below the cmc of SDS and that SDS reduces the size of Pluronic micelles. Thus, at low SDS concentrations, the mixed micelles formed have a small association number with a SDS/EPE molar ratio of about 3–4. The mixed micelles are likely spherical in shape with the coiled POP blocks solubilized in the interior of the SDS micelle.

The Bayreuth group studied the influence of SDS on the association and gelation of Pluronic F-127 using a variety of experimental methods [14–16]. SDS binds to Pluronic F-127 strongly and can suppress the micelle formation of Pluronic F-127 completely. At saturation, about 4–6 SDS molecules bind to one Pluronic F-127 molecule. The Pluronic F-127/SDS complex seems to be stretched in the POP region, as the SDS molecule is presumably bound to the hydrophobic PO group. In connection with the suppression of polymeric micelles, the gel region of Pluronic F-127 decreases with SDS concentration and finally vanishes completely.

The Pluronic L-64/water system forms anisotropic liquid crystalline phases (i.e., hexagonal and lamellar phases) at high copolymer concentrations (Fig. 13). The addition of SDS leads to the transformation from a anisotropic mesophase into an isotropic solution or from a bicontinuous isotropic solution into a solution consisting of discrete micelles [61].

VII. MICELLE FORMATION AND GELATION OF EO/BO POLYMERIC SURFACTANTS

The hydrophobic part of Pluronic polymeric surfactants can be replaced by a more lipophilic hydrophobe [e.g., the polyoxybutylene (POB) block] to form another class of amphiphilic block copolymers. Some early results on their association and gelling properties were reported as patents [140]. The compounds became commercially available only recently and are now manufactured by the Dow Chemical Co. The anionic polymerization of propylene oxide is more or less accompanied by a side reaction (transfer reaction) which finally leads to the formation of diblock copolymer [73]. Thus, the appearance of a few percent of diblock copolymers as composition contaminant is not unexpected for commercial Pluronic products. On the other hand, this kind of side reaction is absent with the anionic polymerization of butylene oxide; therefore the EBE block copolymers produced can be expected to be free of this disadvantage.

In the past few years the Manchester research group has completed extensive studies on the association and gelation properties of E/B block copolymers in aqueous medium, including the EB, EBE, and BEB series [72,81–83,85–90,92]. The solution properties of the two classes of copolymer surfactants (E/B and E/P) are similar in many aspects. For instance, the POB block is mainly responsible for the formation of micelles and, accordingly, the cmc decreases with increasing temperature. Under the same conditions, as the POB block length increases, the cmc decreases but the association number increases. There

TABLE 10 Critical Micelle Concentration of Aqueous Solutions of EB, EBE, and BEB Block Copolymers

Copolymer composition	mw	Temp. (°C)	cmc (g dm^{-3})	Method[a]	Ref.
$E_{50}B_4$	2500	40	≥ 45	ST	85
$E_{27}B_5$	1550	40	≥ 8	ST	85
$E_{41}B_8$	2400	40	3×10^{-1}	ST	90
$E_{24}B_{10}$	1750	40	2.3×10^{-2}	ST	85
$E_{50}B_{13}$	3100	20	3×10^{-2}	ST	85
		40	1×10^{-2}	ST	85
$E_{21}B_8E_{21}$	2400	20	23	LS	90
		30	7	LS	90
		40	2	LS	90
$E_{40}B_{15}E_{40}$	4700	21	2×10^{-1}	LS	83
		37	1.3×10^{-1}	LS	83
$E_{58}B_{17}E_{58}$	6400	30	3.0×10^{-1}	LS, ST	81
		40	1.5×10^{-1}	ST	81
		40	2.0×10^{-1}	LS	81
$E_{71}B_{28}E_{71}$	8000	30	1.6×10^{-1}	LS	82
$E_{132}B_{53}E_{132}$	15,500	30	$\leq 2 \times 10^{-2}$	LS	82
$B_4E_{40}B_4$	2350	40	35	LS	90
		50	16	LS	90

[a] ST, surface tension; LS, light scattering.

is one important difference; because oxybutylene is more hydrophobic than oxypropylene, the same level of association and gelation properties could be achieved at smaller POB block length and higher POE content for the E/B block copolymers. In other words, a larger surface activity and a lower cmc could be predicted. Table 10 summarizes the existing cmc data for this class of polymeric surfactants.

Based on published results obtained by a combination of various techniques, several general remarks can be made about the association and gelation characteristics of the E/B copolymer surfactants in aqueous medium:

1. From the relationship between the standard free energy of micellization and the number of BO units of E_mB_n diblock copolymers, the contribution to ΔG^0 of each BO unit is about −2.5 kJ/mol at 40 °C [85] similar to the value of −2.6 kJ/mol for a CH_2 group of the normal nonionic surfactant

$C_m(EO)_n$. In this connection, one BO unit is equivalent to one CH_2 group or to four PO units.

2. For E/B diblock copolymers, it is necessary to use a minimum of 4 BO units as the hydrophobe [87] in order for the micellization to be carried out at practically accessible concentrations (Table 10). As noted before, 10–13 PO units are for EPE polymeric surfactants. Also for $E_{30}B_n$, gelation has been observed only when n is equal to 6–7 or more [87].
3. The chain architecture has great influence on the selfassembly ability and the gelation tendency. For block copolymers with the same composition (e.g., $N_{BO} = 8$ and $N_{EO} = 40$–42), the cmc values are in the following order [90]:

   EB diblock < EBE triblock < BEB triblock

   Also, the association number and the micelle size follow the following sequence:

   BEB triblock < EBE triblock ≤ EB diblock

   and the clouding point behaves as

   EB diblock ≥ EBE triblock > BEB triblock

   The order for the gelation tendency is

   EB diblock ≥ EBE triblock > BEB triblock

4. For most of the EB diblocks and EBE triblocks studied, the critical gel concentration (cgc) (i.e., the minimum copolymer concentration required to form a gel) is in the range 12–30 wt% [86], depending on the POB content and the total molecular weight. The cgc data are well correlated with the micellar molar mass and the equivalent hard sphere radius of the micelle, thus supporting the close packing of hard spheres as the gelation mechanism [83,86,90,97]. However, the diblock copolymer $E_{27}B_7$ reveals a sol-gel-sol transition by varying the concentration from 40 to 60 wt% at constant temperature. This is likely associated with the change in micellar structure at high concentrations, but needs to be proved.
5. For $E_{40}B_{10}$ and $E_{41}B_8$, the concentration change at the upper gel-sol boundary with the amount of salt added (≤ 0.32 M K_2SO_4) can be well correlated with the excluded volume change of micelles when interpreted by the close packing of hard spheres [92].

VIII. THEORETICAL DEVELOPMENT

Theoretical treatments of block copolymer micelles have been pioneered by de Gennes [163], Leibler, Orland and Wheeler [164], and Noolandi and coworkers

[165,166]. Nagarajan and Ganesh [123] developed a theory of micelle formation of diblock copolymers in selective solvents whereby the corona block (POE) interacts favorably with the solvent (H_2O). The empirical scaling relation, core radius, and corona thickness could be obtained by fitting numerical computational results. They also made a theoretical treatment of solubilization in block copolymer micelles by using a uniform stretching model and a nonuniform stretching model for the chain deformation in the core and corona regions. The scaling analyses have been extended on an analogy between micelles and star polymers [167].

In the past several years, Linse [49,54–57] has made extensive studies on temperature-dependent micellization [49,55], phase behavior [54], micellization in the presence of polydispersity [56], and impurity effects of poly(ethylene oxide-co-propylene oxide) block copolymers [57] in aqueous solution. Many of the theoretical results are in agreement with experiments. Other relevant theoretical work includes that by Hatton and coworkers [64,65] and by Price and coworkers [75], as well as the simulations by Mattice and coworkers [125,126].

ACKNOWLEDGEMENTS

BC gratefully acknowledges the support of this work by the Polymers Program, National Science Foundation (DMR 9301294), the Department of Energy (DEFG0286ER45237.011) and the U.S. Army Research Office (DAAH0494G0053). He also wishes to thank the Human Frontier Science Program for the generous support on work related to gels and to many of his colleagues and fellow scientists who have graciously sent him their reprints and preprints; in particular, to C. Booth, W. Brown, W-D. Hergeth, H. Hoffmann, B. Lindman, P. Linse, M. Malmsten, W. L. Mattice, K. Mortensen, R. Nagarajan, and C. Price.

REFERENCES

1. I. R. Schmolka, in *Polymers for Controlled Drug Delivery* (P. J. Tarcha, Ed.), CRC Press, Boston, 1991, Chapter 10.
2. See also I. R. Schmolka, *J. Amer. Oil Chem. Soc. 54*:110 (1977).
3. Z. Tuzar and P. Kratochvil, in *Surface and Colloid Science* (E. Matijevic, Ed.), Plenum Press, New York, 1993, Vol. 15, pp. 1–83.
4. I. R. Schmolka, in *Nonionic Surfactants* (M. J. Schick, Ed.), Surfactant Science Series, Vol. 1, Marcel Dekker, New York, 1967; Chapter 10.
5. B. Chu, *Langmuir 11*:414 (1995).
6. N. M. Mitchard, A. E. Beezer, N. H. Rees, J. C. Mitchell, S. Leharne, B. Z. Chowdhry, and G. Buckton, *J. Chem. Soc., Chem. Commun.* 900 (1990).

7. A. E. Beezer, J. C. Mitchell, N. H. Rees, J. K. Armstrong, B. Z. Chowdhry, S. Leharne, and G. Buckton, *J. Chem. Research (S)*, 254 (1991).
8. A. E. Beezer, N. M. Mitchard, J. C. Mitchell, J. K. Armstrong, B. Z. Chowdhry, S. A. Leharne, and G. Buckton, *J. Chem. Research (S)*, 236 (1992).
9. N. M. Mitchard, A. E. Beezer, J. C. Mitchell, J. K. Armstrong, B. Z. Chowdhry, and S. Leharne, *J. Phys. Chem. 96*:9507 (1992).
10. J. K. Armstrong, J. Parsonage, B. Z. Chowdhry, S. Leharne, J. C. Mitchell, A. E. Beezer, and P. Laggner, *J. Phys. Chem. 97*:3904 (1993).
11. A. E. Beezer, W. Loh, J. C. Mitchell, P. G. Royall, D. O. Smith, M. S. Tute, J. K. Armstrong, B. Z. Chowdgry, S. A. Leharne, D. Eagland, and N. J. Crowther, *Langmuir 10*:4001 (1994).
12. G. Wanka, H. Hoffmann, and W. Ulbricht, *Colloid Polym. Sci. 268*:101 (1990).
13. G. Wanka, H. Hoffmann, and W. Ulbricht, *Macromolecules 27*:4145 (1994).
14. E. Hecht and H. Hoffmann, *Langmuir 10*:86 (1994).
15. E. Hecht and H. Hoffmann, *Colloids Surf. A96*:181 (1995).
16. E. Hecht, K. Mortensen, M. Gradzielski, and H. Hoffmann, *J. Phys. Chem. 99*:4866 (1995).
17. W. Brown, K. Schillen, M. Almgren, S. Hvidt, and P. Bahadur, *J. Phys. Chem. 95*:1850 (1991).
18. W. Brown, K. Schillen, and S. Hvidt, *J. Phys. Chem. 96*:6038 (1992).
19. P. Lianos and W. Brown, *J. Phys. Chem. 26*:6439 (1992).
20. K. Schillen, O. Glatter, and W. Brown, *Progr. Colloid Polym. Sci. 93*:66 (1993).
21. K. Schillen, W. Brown, and C. Konak, *Macromolecules 26*:3611 (1993).
22. K. Schillen, W. Brown, and R. M. Johnsen, *Macromolecules 27*:4825 (1994).
23. O. Glatter, G. Scherf, K. Schillen, and W. Brown, *Macromolecules, 27*:6046 (1994).
24. S. Hvidt, E. B. Jorgenson, W. Brown, and K. Schillen, *J. Phys. Chem. 98*:12320 (1994).
25. M. Almgren, J. Alsins, and P. Bahadur, *Langmuir 7*:446 (1991).
26. M. Almgren, J. Stam, C. Lindblad, P. Li, P. Stilbs, and P. Bahadur, *J. Phys. Chem. 95*:5677 (1991).
27. M. Almgren, P. Bahadur, M. Jansson, P. Li, W. Brown, and A. Bahadur, *J. Colloid Interface Sci. 151*:157 (1992).
28. M. Almgren, W. Brown, and S. Hvidt, *Colloid Polym. Sci. 273*:2 (1995).
29. P. Bahadur, P. Li, M. Almgren, and W. Brown, *Langmuir 8*:1903 (1992).
30. P. Bahadur and K. Pandya, *Langmuir 8*:2666 (1992).

31. P. Bahadur, K Pandya, M. Almgren, P. Li, and P. Stilbs, *Colloid Polym. Sci. 271*:657 (1993).
32. K. Pandya, K. Lad, and P. Bahadur, *J. Macromol. Sci.-Pure Appl. Chem. A30*:1 (1993).
33. K. Pandya, P. Bahadur, T. N. Nagar, T. N., and A. Bahadur, *Colloids Surf. A70*:219 (1993).
34. K. Mortensen, W. Brown, and B. Norden, *Phys. Rev. Letts. 68*:2340 (1992).
35. K. Mortensen, *Europhys. Letters 19*:599 (1992).
36. K. Mortensen and J. K. Pedersen, *Macromolecules 26*:805 (1993).
37. K. Mortensen and W. Brown, *Macromolecules 26*:4128 (1993).
38. K. Mortensen, *Progr. Colloid Polym. Sci. 91*:69 (1993).
39. K. Mortensen, *Progr. Colloid Polym. Sci. 93*:72 (1993).
40. K. Mortensen, D. Schwahn, and S. Janssen, *Phys. Rev. Letts. 71*:1728 (1993).
41. K. Mortensen, *J. Phys. IV 3(C8)*:157 (1993).
42. K. Mortensen, W. Brown, and E. Jorgensen, *Macromolecules 27*:5654 (1994).
43. K. Mortensen, W. Brown, and E. Jorgensen, *Macromolecules 28*:1458 (1995).
44. A. A. Samii, B. Lindman, and G. Karlstrom, *Progr. Colloid Polym. Sci. 82*:280 (1990).
45. A. A. Samii, G. Karlstrom, and B. Lindman, *Langmuir 7*:1067 (1991).
46. F. Tiberg, M. Malmsten, P. Linse, and B. Lindman, *Langmuir 7*:2723 (1991).
47. M. Malmsten, P. Linse, and T. Cosgrove, *Macromolecules 25*:2474 (1992).
48. G. Karlstrom and B. Lindman, in *Organized Solutions. Surfactants in Science and Technology* (S. E. Friberg and B. Lindman, Eds.), Marcel Dekker, New York, 1992, Chapter 5.
49. P. Linse and M. Malmsten, *Macromolecules 25*:5434 (1992).
50. M. Malmsten and B. Lindman, *Macromolecules 25*:5440 (1992).
51. M. Malmsten and B. Lindman, *Macromolecules 25*:5446 (1992).
52. M. Malmsten and B. Lindman, *Macromolecules 26*:1282 (1993).
53. M. Malmsten, P. Linse, and K-W. Zhang, *Macromolecules 26*:2905 (1993).
54. P. Linse, *J. Phys. Chem. 97*:13896 (1993).
55. P. Linse, *Macromolecules 26*:4437 (1993).
56. P. Linse, *Macromolecules 27*:6404 (1994).
57. P. Linse, *Macromolecules 27*:2685 (1994).
58. M. H. G. M. Penders, S. Nilsson, L. Piculell, and B. Lindman, *J. Phys. Chem. 98*:5508 (1994).

59. K. Zhang, M. Carlsson, P. Linse, and B. Lindman, *J. Phys. Chem. 98*:2542 (1994).
60. K. Zhang, Ph.D. Thesis, University of Lund, 1994.
61. K. Zhang, B. Lindman, and L. Coppola, *Langmuir 11*:538 (1995).
62. K. Zhang and M. Carlsson, *J. Phys. Chem. 99*:5051 (1995).
63. P. N. Hurter and T. A. Hatton, *Langmuir 8*:1291 (1992).
64. P. N. Hurter, M. H. M. Scheutjens, and T. A. Hatton, *Macromolecules 26*:5592 (1993).
65. P. N. Hurter, M. H. M. Scheutjens, and T. A. Hatton, *Macromolecules 26*:5030 (1993).
66. P. Alexandridus, J. F. Holzwarth, and T. A. Hatton, *Macromolecules 27*:2414 (1994).
67. P. Alexandridus, V. Athanassiou, S. Fukuda, and T. A. Hatton, *Langmuir 10*:2604 (1994).
68. J. Rassing and D. Attwood, *Int. J. Pharm. 13*:47 (1983).
69. C. Price, N. Briggs, J. R. Quintana, R. B. Stubbersfield, and I. Robb, *Polym. Commun. 27*:292 (1986).
70. N. K. Reddy, P. J. Fordham, D. Attwood, and C. Booth, *J. Chem. Soc. Faraday Trans. 86*:1569 (1990).
71. N. K. Reddy, A. Foster, M. G. Styring, and C. Booth, *J. Colloid Interface Sci. 136*:588 (1990).
72. W-B. Sun, J-F. Ding, R. H. Mobbs, D. Attwood, and C. Booth, *Colloids Surf. 54*:103 (1991).
73. J-F. Ding, F. Heatley, C. Price, and C. Booth, *Eur. Polym. J. 27*:895 (1991).
74. J-F. Ding, D. Attwood, C. Price, and C. Booth, *Eur. Polym. J. 27*:901 (1991).
75. X-F. Yuan, A. J. Masters, and C. Price, *Macromolecules 25*:6876 (1992).
76. Y-L. Deng, J-F. Ding, R. B. Stubbersfield, F. Heatley, D. Attwood, C. Price, and C. Booth, *Polymer 33*:1963 (1992).
77. L. Yang, A. D. Bedells, D. Attwood, and C. Booth, *J. Chem. Soc. Faraday Trans 88*:1447 (1992).
78. Y-L. Deng, G-E. Yu, C. Price, and C. Booth, *J. Chem. Soc. Faraday Trans. 88*:1441 (1992).
79. Q-G. Wang, C. Price, and C. Booth, *J. Chem. Soc. Faraday Trans. 88*:1437 (1992).
80. G-E. Yu, Y-L. Deng, S. Dalton, Q-G. Wang, D. Attwood, C. Price, and C. Booth, *J. Chem. Soc. Faraday Trans. 88*:2537 (1992).
81. Y-Z. Luo, C. V. Nicholas, D. Attwood, J. H. Collett, C. Price, and C. Booth, *Colloid Polym. Sci. 270*:1094 (1992).
82. C. V. Nicholas, Y-Z. Luo, N-J. Deng, D. Attwood, J. H. Collett, C. Price, and C. Booth, *Polymer 34*:138 (1993).

83. Y-Z. Luo, C. V. Nicholas, D. Attwood, J. H. Collett, C. Price, C. Booth, B. Chu, and Z-K. Zhou, *J. Chem. Soc. Faraday Trans. 89*:539 (1993).
84. Q-G. Wang, G-E. Yu, Y-L. Deng, C. Price, and C. Booth, *Eur. Polym. J. 29*:665 (1993).
85. A. D. Bedells, R. M. Arafeh, Z. Yang, D. Attwood, F. Heatly, J. C. Padget, C. Price, and C. Booth, *J. Chem. Soc. Faraday Trans. 89*:1235 (1993).
86. A. D. Bedells, R. M. Arafeh, Z. Yang, D. Attwood, J. C. Padget, C. Price, and C. Booth, *J. Chem. Soc. Faraday Trans. 89*:1243 (1993).
87. S. Tanodekaew, N-J. Deng, S. Smith, Y-W. Yang, D. Attwood, and C. Booth, *J. Phys. Chem. 97*:11847 (1993).
88. Z-G. Yan, Z. Yang, C. Price, and C. Booth, *Makromol. Chem. Rapid. Commun. 14*:725 (1993).
89. Y-L. Deng, C. Price, and C. Booth, *Eur. Polym. J. 30*:103 (1994).
90. Z. Yang, S. Pickard, N-J. Deng, R. J. Barlow, D. Attwood, and C. Booth, *Macromolecules 27*:2371 (1994).
91. P. R. Knowles, R. J. Barlow, F. Heatley, C. Booth, and C. Price, *Makromol. Chem. Phys. 195*:2547 (1994).
92. N-J. Deng, Y-Z. Luo, S. Tanodekaew, N. Bingham, D. Attwood, and C. Booth, *J. Polym. Sci. Polym. Phys. Ed. 33*:1085 (1995).
93. Q-G. Wang, M. Savage, M. Draper, J. H. Collett, D. Attwood, C. Price, and C. Booth, *J. Pharm. Res.* submitted.
94. I. Alig, R-V. Ebert, W-D. Hergeth, and S. Wartewig, *Polym. Commun. 31*:314 (1990).
95. S. Wartewig, I. Alig, W-D. Hergeth, J. Lange, R. Lochmann, and T. Scherzer, *J. Mol. Struct. 219*:365 (1990).
96. W-D. Hergeth, I. Alig, J. Lange, J. R. Lochmann, T. Scherzer, and S. Wartewig, *Makromol. Chem., Macromol. Symp. 52*:289 (1991).
97. W-D. Hergeth, R. Zimmermann, P. Bloss, K. Schmutzler, and S. Wartewig, *Colloids Surf. 56*:177 (1991).
98. P. Bloss, W-D. Hergeth, C. Wohlfarth, and S. Wartewig, *Makromol. Chem. 193*:957 (1992).
99. G. Fleischer, P. Bloss, and W-D. Hergeth, *Colloid Polym. Sci. 271*:217 (1993).
100. G. Fleischer, *J. Phys. Chem. 97*:517 (1993).
101. Z. Zhou and B. Chu, *Macromolecules 20*:3089 (1987).
102. Z. Zhou and B. Chu, *Macromolecules 21*:2548 (1988).
103. Z. Zhou and B. Chu, *J. Colloid Interface Sci. 126*:171 (1988).
104. A. Tontisakis, R. Hilfiker, and B. Chu, *J. Colloid Interface Sci. 135*:427 (1990).
105. G. Wu, Z. Zhou, and B. Chu, *Macromolecules 26*:2117 (1993).
106. G. Wu, Z. Zhou, and B. Chu, *J. Polym. Sci. Polym. Phys. Ed. 31*:2035 (1993).

107. Z. Zhou, B. Chu, and D. G. Peiffer, *Macromolecules 26*:1876 (1993).
108. G. Wu and B. Chu, *Macromolecules 27*:1766 (1994).
109. Z. Zhou and B. Chu, *Macromolecules 27*:2025 (1994).
110. B. Chu, Z. Zhou, and G. Wu, *J. Non-Cryst. Solids 172–174*:1094 (1994).
111. Z. Zhou, B. Chu, and D. G. Peiffer, *J. Polym. Sci. Polym. Phys. Ed. 32*:2135 (1994).
112. G. Wu, B. Chu, and D. K. Schneider, *J. Phys. Chem. 98*:12018 (1994).
113. B. Chu, G. Wu, and D. K. Schneider, *J. Polym. Sci. Polym. Phys. Ed. 32*:2605 (1994).
114. G. Wu, Q. Ying, and B. Chu, *Macromolecules 27*:5758 (1994).
115. Z. Zhou, B. Chu, and D. K. Peiffer, *Langmuir 11*:1956 (1995).
116. G. Wu and B. Chu, *Macromol. Symp. 90*:251 (1995).
117. G. Wu, Ph.D. Thesis, State University of New York at Stony Brook, 1994.
118. N. J. Turro and C. Chung, *Macromolecules 17*:2123 (1984).
119. N. J. Turro and P-L. Kuo, *Langmuir 2*:438 (1986).
120. N. J. Turro and P-L. Kuo, *J. Phys. Chem. 90*:4205 (1986).
121. N. J. Turro and P-L. Kuo, *Langmuir 3*:773 (1987).
122. R. Nagarajan, M. Barry, and E. Ruckenstein, *Langmuir 2*:210 (1986).
123. R. Nagarajan and K. Ganesh, *J. Chem. Phys. 90*:5843 (1989).
124. R. Nagarajan and K. Ganesh, *Macromolecules 22*:4312 (1989).
125. Y. Wang, W. L. Mattice, and D. H. Napper, *Macromolecules 25*:4073 (1992).
126. K. Rodrigues and W. L. Mattice, *Langmuir 8*:456 (1992).
127. K. N. Prasad, T. T. Luong, A. T. Florence, J. Paris, C. Vaution, M. Seiller, and F. Puisieux, *J. Collid Interface Sci. 69*:225 (1979).
128. J. B. Kayes and D. A. Rawlins, *Colloid Polym. Sci. 257*:622 (1979).
129. T. F. Tadros and B. Vincent, *J. Phys. Chem. 84*:1575 (1980).
130. A. A. Al-Saden, T. L. Whateley, and A. T. Florence, *J. Colloid Interface Sci. 90*:303 (1982).
131. J. Rassing, W. P. McKenna, S. Bandyopadhyay, and E. M. Eyring, *J. Mol. Liquids 27*:165 (1984).
132. M. Vadnere, G. Amidon, S. Lindenbaum, and J. L. Haslam, *Int. J. Pharm. 22*:207 (1984).
133. R. K. Williams, M. A. Simard, and C. Jolicoeur, *J. Phys. Chem. 89*:178 (1985).
134. E. Killmann, H. Maier, and J. A. Baker, *Colloids Surf. 31*:51 (1988).
135. J. A. Baker and J. C. Berg, *Langmuir 4*:1055 (1988).
136. J. A. Baker, R. A. Pearson, and J. C. Berg, *Langmuir 5*:339 (1989).
137. J. Juhasz, V. Lenaerts, P. V. M. Tan, and H. Ong, *J. Colloid Interface Sci. 136*:168 (1990).
138. L. M. A. van de Steeg and C-G. Golander, *Colloids Surf. 55*:105 (1991).
139. I. R. Schmolka and L. Bacon, *J. Amer. Oil Chemist Soc. 44*:559 (1967).

140. I. R. Schmolka, *J. Amer. Oil Chemist Soc. 68*:206 (1991).
141. P. Wang and T. P. Johnston, *J. Appl. Polym. Sci. 43*:283 (1991).
142. C. Camire, L. Meilleur, and F. Quirion, *J. Phys. Chem. 96*:2360 (1992).
143. K. Nakashima, T. Anzai, and Y. Fujimoto, *Langmuir 10*:658 (1994).
144. B. Chu, Laser Light Scattering. *Basic Principles and Practice, 2nd Edition*, Academic Press, New York, 1991.
145. T. P. Russell, In *Handbook on Synchrotron Radiation, Vol. 3*, (G. S. Brown and D. E. Moncton, Eds.), North Holland, Amsterdam, 1991, Chapter 11, pp. 379–469.
146. O. Glatter and O. Kratky (eds.), *Small Angle X-ray Scattering*, Academic Press, New York, 1982.
147. K. Meguro, M. Ueno, and K. Esumi, In *Nonionic Surfactants. Physical Chemistry* (M. J. Schick, Ed.), Surfactant Science Series, Vol. 23, Marcel Dekker, New York, 1987, Chapter 3.
148. P. Mukerjee and K. J. Mysels, *Critical Micelle Concentrations of Aqueous Surfactant System*, NSRDS-NBS 36, Goverment Printing Office, Washington, D.C., 1971.
149. T. Kotaka, N. Donkai, and T. Ik, *Min. Bull. Inst. Chem. Res. Kyoto Univ. 52*:232 (1974).
150. A. Sikora and Z. Tuzar, *Makromol. Chem. 184*:2049 (1983).
151. S. W. Provencher, *Makromol. Chem. 180*:201 (1979).
152. S. W. Provencher, *Comput. Phys. Commun. 27*:213, 229 (1982).
153. G. ten Brinke and G. Hadziioannou, *Macromolecules 20*:486 (1987).
154. L. A. Girifilco and R. J. Good, *J. Phys. Chem. 61*:904 (1957).
155. E. H. Crook, D. B. Fordyce, and G. F. Trebbi, *J. Phys. Chem. 67*:1987 (1963).
156. M. J. Rosen, *Surfactants and Interfacial Phenomena, 2nd Edition*, J. Wiley & Sons, New York, 1989.
157. J. Sjoblom, P Stenius, and I. Danielsson, In *Nonionic Surfactants. Physical Chemistry* (M. J. Schick, Ed.), Surfactant Science Series, Vol. 23, Marcel Dekker, New York, 1987; Chapter 7.
158. D. J. Mitchell, G. J. T. Tiddy, L. Waring, T. Bostock, and M. P. MacDonald, *J. Chem. Soc. Faraday Trans. I 79*:975 (1983).
159. J. K. Percus and G. J. Yevick, *Phys. Rev. 110*:1 (1958).
160. P. N. Pusey and W. van Megen, *Nature 320*:340 (1986).
161. BASF Corporation Brochure on Pluronic and Tetronic Surfactants, 1989.
162. J. S. Higgins and H. C. Benoit, *Polymers and Neutron Scattering*, Oxford University Press Inc., New York, 1994.
163. P. G. de Gennes, In *Solid State Physics* (J. Liebert, Ed.), Academic Press, New York, 1978; Suppl. 14, p. 1.
164. L. Leibler, H. Orland, and J. C. Wheeler, *J. Chem. Phys. 79*:3550 (1983).
165. J. Noolandi and M. H. Hong, *Macromolecules 16*:1443 (1983).

166. D. Whitemore and J. Noolandi, *Macromolecules 18*:657 (1985).
167. R. Nagarajan and K. Ganesh, *J. Chem. Phys. 98*:7440 (1993).
168. X. Peng and Z. Zhou, *Acta Chimica Sinica 44*:613 (1986).
169. L. Dai, P. Wu, and Z. Zhou, *Acta Physico-Chimica Sinica 8*:22 (1992).

4

Properties of Polyoxyalkylene Block Copolymers

VAUGHN M. NACE Industrial Polyglycol Research and Development, The Dow Chemical Company, Freeport, Texas

I. Introduction 146
A. Property categories 146
B. Nomenclature 147

II. Physical Properties of Polyoxyalkylene Block Copolymers 147
A. Physical state 147
B. Melt viscosity characteristics 154
C. Polyoxyalkylene polarity 156
D. Chemical and thermal stability 161

III. Performance-Oriented Properties 163
A. Wetting 164
B. Foaming 170
C. Comparisons of copolymers based on POP and POB hydrophobes 173

References 178

I. INTRODUCTION

A. Property Categories

For any class of surfactant we would like to define a set of criteria which can be used to differentiate relative performance for a given application. Ideally, the researcher and product formulation chemist draw information from a vast set of available performance data and make logical decisions as to the surfactant or surfactant class to choose for experimentation. In many of these instances, however, the surfactant of interest may not even remotely exhibit the right properties and economic realities for the new system under development. Reasons for the divergence from the "cook-book" ideal may be numerous. Some properties may fall in line with desired performance needs, while others could cause problems in material handling and pricing. Price, while not a topic to be covered in this volume, is a very important property and this aspect alone can often times be the final determinant in the noncommercialization of technically impressive research projects.

A reasonable approach for presenting performance data for a class of surfactants is to divide the information into two categories: 1) physical properties, and 2) performance-oriented properties. It is not surprising that these property classes interact in many ways and, depending on the intended application, interactions can be very complex and unpredictable. For example, Rosen and Zhu [1] reported unexpected synergistic wetting effects of water soluble/insoluble surfactant pairs. Other relationships between property classes will come to light as the chapter unfolds. Therefore, there can be no real, absolute line of demarcation between these two arbitrary property categories, although they are useful for constructing a coherent foundation for discussion.

Physical properties include, but are not limited to general attributes such as solid/liquid behavior, viscosity, hydrophobe polarity, solubility, and chemical stability. Polarity differences between POP and POB hydrophobes and comparisons to other common hydrophobes will be covered in detail.

Performance-oriented properties are meaningful in the context of specific processes such as detergency, wetting, foaming, solubilization, dispersion, and emulsification. Phase behavior is also related to performance, especially when speaking of gelation and cloud point. Studies on the topics of detergency, phase behavior, emulsification, dispersion, and solubilization are fairly numerous, but will not be covered in this chapter owing to the wide degree of variability in test methods and substrates. Interfacial performance properties are important when considering the complex nature of phase interfaces (liquid/gas, liquid/liquid, and liquid/solid being the most important). Moreover, the nature of the interface is related to all performance-oriented properties. Surface and interfacial tension analysis provides information about the efficiency and effectiveness of a surfactant, and, to some degree, the interfacial architecture (e.g., the area per molecule at the interface). Another important figure of merit includes the C_{20}, which is the concentration at which the interface is effectively saturated with surfactant monomer. One well-studied interfacial property is the critical micellization concentration (cmc). Interfacial property data covered here will be mostly limited to the comparison of E-P-E and E-B-E block copolymers (E, polyoxyethylene; P, polyoxypropylene; B, polyoxybutylene). A thorough discussion of more detailed interfacial properties, such as phase behavior, cloud point, micellization, and gel behavior, will be found in the chapter on physical chemistry.

Finally, one should realize that surfactant structure is the key variable that ultimately determines all properties. Other more subtle relationships exist between synthetic conditions and various surfactant properties. Because these relationships are not taken into account in many studies on block copolymer surfactants, they are covered in detail in the chapter on polyoxyalkylene block copolymer synthesis.

B. Nomenclature

The following nomenclature in this chapter will be used consistently. The general term of polyoxyalkylene (POA) block copolymers will be used throughout. Polyoxyethylene (POE), polyoxypropylene (POP), and polyoxybutylene (POB) will be used to refer to homopolymers and homopolymeric blocks made from ethylene oxide (EO), propylene oxide (PO), and 1,2–butylene oxide (BO). For describing block structures, a short-hand notation will be used; for example, a POE-POP-POE triblock copolymer having hydroxyl groups at each end will be written as E-P-E. If end groups other than hydroxyl are present, they will be specified.

II. PHYSICAL PROPERTIES OF POLYOXYALKYLENE BLOCK COPOLYMERS

A. Physical State

The physical behavior of a nonionic surfactant becomes important when material handling issues arise. Solid/liquid behavior and viscosity are two factors which must be understood in terms of surfactant structure.

Figure 1 shows the simplified triblock structure of a typical POA block copolymer. In this example, the center block consists of a POB or POP hydrophobe. Flanking the hydrophobe are two hydrophilic POE blocks. The values of X, Y, and Z relate to the average molecular weight of the material, and therefore the viscosity. The crystalline behavior (and thus the melting behavior) is determined primarily by the average degree of ethoxylation of the POE blocks.

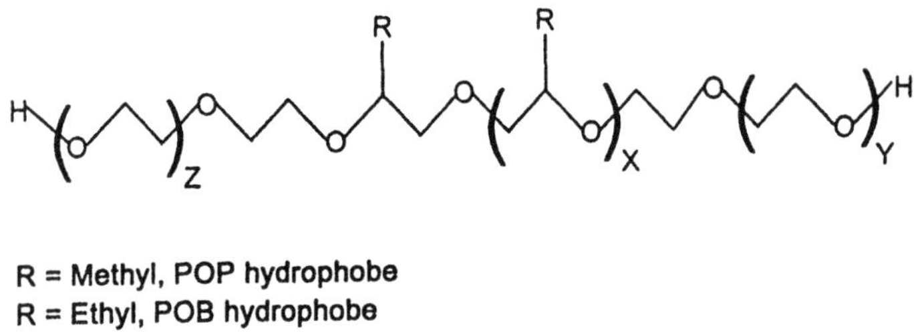

R = Methyl, POP hydrophobe
R = Ethyl, POB hydrophobe

FIG. 1 Simplified structure of triblock copolymer of the type E-B-E or E-P-E.

POP and POB blocks found in base-catalyzed commercial copolymers are generally amorphous and do not effect crystallinity to a large degree [2,3]. Therefore, these surfactants consist of a noncrystalline hydrophobic moiety connected to hydrophilic, crystallizable POE chains. Also, depending on the block structure, the hydrophobe or hydrophile may have a terminal hydroxyl group present. POE blocks having fewer than 12–13 EO units are typically noncrystalline at room temperature and are miscible with the hydrophobic portion of the polymer, exhibiting water-like clarity. As the number of EO units in the block increases, the consistency of the surfactant will change to a hazy opaque liquid, then to a paste, and finally to a solid. Before presenting data on crystallization behavior of block copolymers, it is instructional to build a background from studies on simple POE materials and their associated melting behavior.

1. Crystallization and Melting Behavior of POE Diols and Derivatives

Several studies on polyoxyethylene diols have shown the melting point to increase with the degree of ethoxylation, and therefore, with the concomitant decrease in hydroxyl end groups. Crystalline form is primarily dependant on polymer molecular weight, polymer dispersity, and temperature of crystallization. Bailey and Koleske have adequately reviewed the crystallization behavior of POE [4]. Structural features of crystalline POE have been reviewed by Tadokoro [5]. Low-angle X-ray techniques and dilatometry were used by Arlie et al. and Spegt [6,7] to show evidence for terminal hydroxyl groups being located in narrow amorphous layers separating crystalline folded-chain lamellae (Fig. 2). Polymers studied were of relatively high molecular weights ranging from 2,000 to 30,000. The model in Fig. 2 shows the amorphous layer being comprised of end chains (containing hydroxyl end groups), polymer fold points, and transverse polymer sections. Beech et al. [8,9] and Ashman and Booth [10] have done additional work on POE melting behavior by applying Flory and Flory-Vrij theory [11,12]. Estimates of end interfacial free energies of folded and extended chain crystals were made from direct measurements of melting

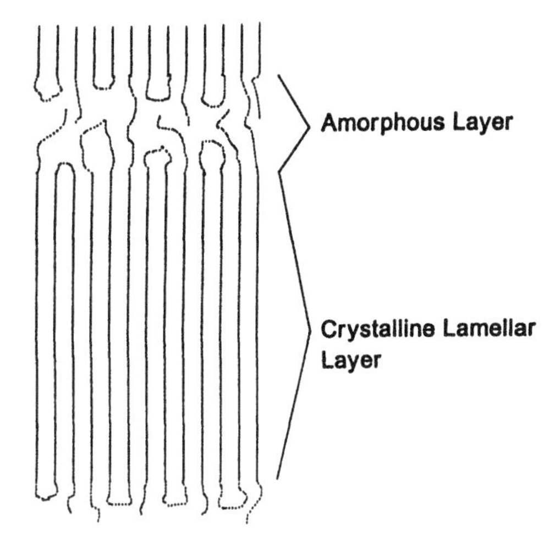

FIG. 2 Model of POE block copolymer showing crystalline and amorphous regions. *Source*: adapted from [6].

point and lamellar dimension. Two important assumptions made by Flory suggest polymer chain ends are excluded from the crystalline matrix, and crystalline/ amorphous boundaries are distinguishable under equilibrium conditions.

Crystallization kinetics information on POE materials with molecular weights ranging from 300 to over 6 million were reported by Hay et al. [13]. Effects of polydispersity differences on crystal growth rates were identified. The thermodynamic melting point ($T_m°$) of an infinitely long POE polymer was determined to be 69 °C. Beech and Booth [14] have determined the value to be 76 °C using fractions of high molecular weight POE. Simek et al. [15] also reported melting data for PEO diols and, by extrapolating a plot of melting point versus inverse degree of polymerization, a $T_m°$ value of 69.4 °C was obtained. Other values (reported or extrapolated) of $T_m°$ range from 67 to 78 °C [13,16,17].

Replacement of the POE terminal hydroxyl group with acetoxy-, phenoxy-, benzyloxy-, or chloro-groups tends to decrease melting temperature [18,19]. Causes of the effects were attributed to differentiated interlamellar interactions brought about by end-group spatial dimension and chain end interactions. However, when methoxy- or ethoxy- end groups were introduced in place of the hydroxyl end group, the melting point was not depressed due to favorable mixing free energy of methyl and ethyl end groups [20].

Melting point depressions of 11–14 °K were reported in the study of monodisperse methyl capped POE materials [21]. Polydisperse methyl-capped samples did not show melting point depressions as compared to uncapped material. Moreover, infrared spectroscopic analysis [22] showed that the crystalline surface layers of monodisperse POE diols are highly ordered as compared to conventional polydisperse diols having disordered surface layers.

2. Melting and Crystallization Behavior of E-P-E and P-E-P Block Copolymer Surfactants

Melting behavior of the POE fraction in block copolymer surfactants is somewhat different than that of POE homopolymers of equivalent block length. For E-P-E triblock copolymers, one end of the POE block is hydroxyl-terminated while the other is in a transitional environment where an EO monomeric unit attaches to a PO unit. For the "reverse" triblock such as P-E-P, there is only one central POE block with both ends characterized by a POE to POP transition.

An interesting question then arises as to the behavior of POE and POP polymer chains as an intermolecular physical mixture of two separate polymers as opposed to a intramolecular mixture such as the triblock copolymers mentioned above. Booth and Pickles [23] studied the interaction of POE 4000 (4000 molecular weight) diols with POP diols having molecular weights ranging from about 250 to 2000. Flory interaction theory [24], polymer physical parameters, and laboratory-generated miscibility data were discussed. While not touched on directly, it is obvious that the weight percentage of hydroxyl groups increases with decreasing POP chain length. This may give partial explanation of the increased miscibility of lower molecular weight POP diols in POE 4000.

Dilatometric studies by Booth and Dodgson [25] on the melting behavior of E-P-E-type block copolymers have shown that melting point depression, with respect to equal chain length POE homopolymers, is brought about by the central POP block and the corresponding POE/POP chain transitions. The magnitude of the depression approaches a maximum of 4 °C. Geczy [26] reported melting point data on a series of E-P-E and P-E-P copolymers. This work used elaborate data manipulation to develop a calculational model for obtaining melting point data. However, a value of 18 °C for the limiting melting point of POP was reported, therefore not having taken into account the amorphous nature of the atactic POP hydrophobe. Figure 3 shows a plot of selected E-P-E copolymer melting data (from [26]) plotted against calculated average POE block length. Average POE block length has been described by Nace et al. from a study of block copolymer melting behavior [16]. The calculation was performed by simply dividing the molecular weight associated with the EO portion of the molecule by 2 (two EO blocks) and then dividing that result by 44 (the molecular weight of an EO monomer unit). The POE block length was uncorrected for secondary hydroxyl contributions. A reasonably good fit is obtained (r = 0.98) using the equation:

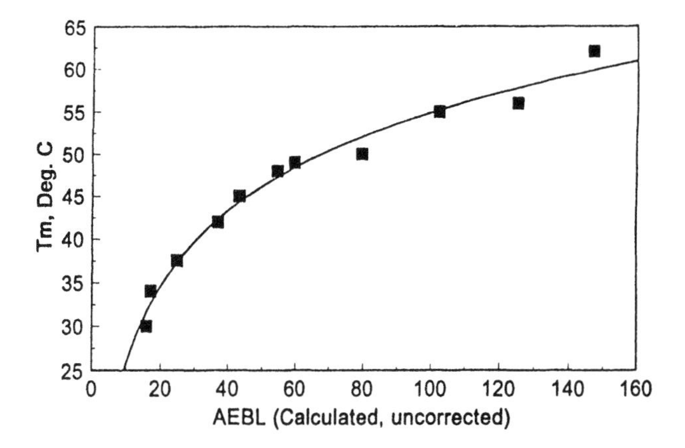

FIG. 3 Melting temperature of E-P-E copolymers versus the calculated average POE block length. *Source*: adapted from [26].

$$T_m = -3.22 + (12.6) \ln (AEBL)$$

where T_m is in °C and *AEBL* (average EO block length) is the average number of EO units per hydrophilic block. Presenting the data in this way, the melting point can be directly related to the portion of the molecule exhibiting crystalline behavior (i.e., the POE hydrophile).

Booth and Pickles conducted a similar study of P-E-P block copolymers using dilatometry and low-angle X-ray scattering [27]. Three sets of copolymers were synthesized, each having central POE blocks of 75, 84, and 98 EO units respectively (3300, 3700, and 4300 mw). The first two sets contained from zero to 13 propylene oxide units per each end, with the third set ranging from zero to 31 PO units. The results showed POE melting point depressions of up to 15 °C, considerably higher when compared to E-P-E copolymers. The magnitude of melting point depression was found to increase with increasing POP chain length. Differences in melting point depression between the two block copolymer types were related to end interfacial free energy and lamellar thickness arguments.

Ashman reported crystallization, melting, and lamellar spacing data on lower molecular weight P-E-P copolymers [28] having a central POE block of 48 EO units flanked with POP blocks each having from zero to 7 PO units. Further study by Viras et al. [29] on a range of P-E-P copolymers showed a solid stacked lamellar phase consisting of tilted chains with respect to the crystal end plane, depending on copolymer structure. Recently, Simek et al. [15] studied melting behavior of P-E-P triblocks having POE block lengths from 8 to 31 units and varying propoxylation levels. Compared to POE diols of similar chain length, melting points of the copolymers were depressed from 8 to 33 °C due to the presence of POP end blocks. They concluded that end interfacial free energy was increased by propoxylation.

Several studies [30–33] on the crystallization of E-P-E and P-E-P block copolymers from dilute ethyl benzene solution demonstrated that monodisperse, monolayer crystals could be obtained using self-seeding techniques. Electrical birefringence, small angle X-ray scattering (SAXS), differential scanning calorimetry (DSC), and gel-permeation chromatography (GPC) were used to study copolymer crystals grown in this fashion.

3. Melting and Crystallization Behavior of P-E Diblock Copolymer Surfactants

Ashman and Booth [34] investigated the melting behavior of methanol-initiated P-E diblock copolymers in which the POE block contained 40 repeat units (1760 mw). Degree of propoxylation ranged from zero to 11. Propoxylation of POE monols depressed the melting point by 3.5–3.8 °C and depended on the crystallization temperature. The crystalline lamellae were approximately 25 EO units in thickness and chain-extended. Amorphous spacing between lamellae ranged from 11.0 nm for the POE monol up to 14.2 nm for an 11-mole propoxylate.

4. Practical Ramifications of Crystallization and Melting Behavior of POA Block Copolymers

In general, commercially-available block copolymers are offered in three distinct physical forms—liquids, pastes, and solids. As for the solids, the final form at room temperature may be prills, flakes, or, in some cases, a cast solid (more aptly described as a solid chunk!) A typical product array of E-P-E materials is sold under the Pluronic® trade name by the BASF Corporation. The polymer nomenclature is indicative of the POP hydrophobe molecular weight, percent EO, and physical state. For instance, the name Pluronic L-62 is indicative of a triblock copolymer with an approximate POP hydrophobe molecular weight of 1800 and 20% by weight of EO spread among the two POE end blocks. The "L" stands for liquid. Descriptors for other Pluronic polymers carry a "P" or "F" prefix which designates paste or flake respectively [35]. At room temperature, Pluronic copolymers having 10–20% EO by weight are typically liquids; those having 30–50% EO are most often pastes, and above 70% EO, they are solids. Thus, one may predict whether triblock copolymers will be pastes or liquids at room temperature by using simple calculation of the average chain length of terminal POE blocks. From the melting data available on POE diols, and knowing the typical melting point depressions caused by end effects of neighboring POP blocks, the following generalizations can be expressed:

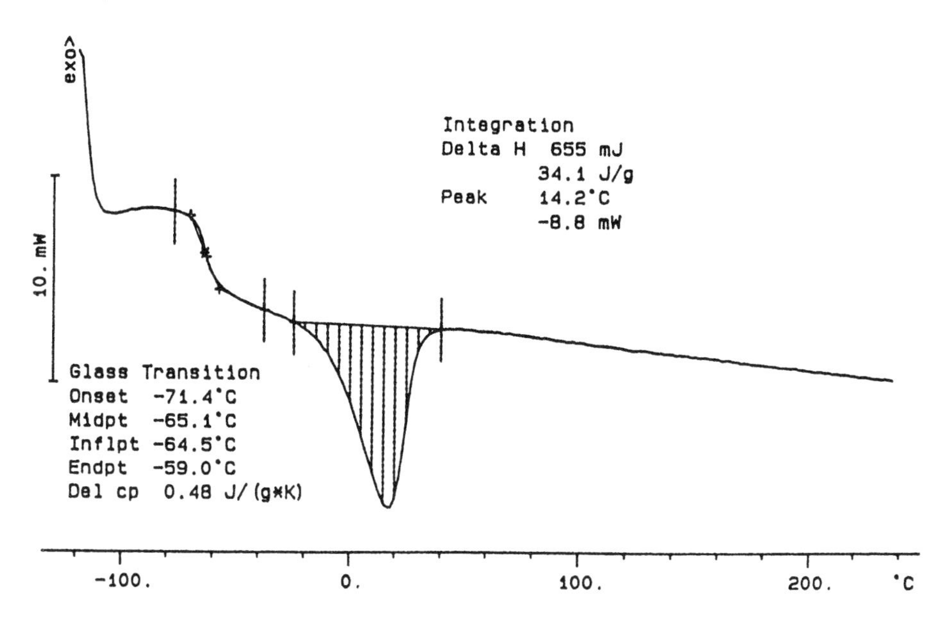

FIG. 4 Second-heat DSC thermogram of a representative E-P-E copolymer surfactant.

1. POE diols are typically clear liquids at room temperature when the chain length is equal to or less than 14–16 EO units.
2. E-P-E triblock copolymers having an average of 16–18 EO units or fewer in the POE end block are liquid at room temperature.
3. P-E-P triblock copolymers with an average POE central block of 23–24 (or fewer) EO units are liquid at room temperature.

Keep in mind that immediately after polymerization block copolymers may be clear at room temperature, but can take on a cloudy appearance after the passage of time due to slow crystallization of higher molecular weight POE blocks present in the normal distribution. Patton [36] showed that random copolymerization of small amounts of PO during EO capping of POP diols leads to E-P-E triblocks having sparkling, stable clarity. This is an example of a process that "breaks up" the crystallinity of the POE blocks and therefore shortens the effective POE block length such that liquidity and an adequate degree of hydrophilicity is maintained.

Differential scanning calorimetry has been effectively used to gain practical information on POA block copolymers. Figure 4 shows a heating thermogram of a representative E-P-E copolymer having a POP-hydrophobe molecular weight of approximately 1800 and capped with 30% EO. At a temperature of approximately 64 °C, a glass transition occurs due to the amorphous POP moiety. At just below room temperature, a melting endotherm transition occurs with a peak temperature of 14.2 °C. From the area above the transition, heat of fusion for the crystalline fraction of the copolymer is obtained, in this case 34.1 J/g. Position of the melting endotherm with respect to temperature can help determine whether the material will contain a crystalline fraction at the use temperature.

B. Melt Viscosity Characteristics

The viscosity of polymers in general, and POA block copolymers in particular, is a function of several variables, the most important of which is degree of polymerization and temperature [37]. For low molecular weight polyethers, the weight percentage of hydroxyl groups plays a very important role in viscosity. In this regard, there can be dramatic discontinuities in viscosity as a function of degree of polymerization for low molecular weight (high percent hydroxyl) species. However, most POA block copolymers useful as surfactants have molecular weights sufficiently high so that hydroxyl end-group effects on viscosity are negligible.

Typical triblock copolymers are made piece-wise. In other words, the hydrophobe (or hydrophile) is made first, then the end-block(s) are added by reacting with a different oxide (or mixture of oxides). Therefore, it is instructional to first briefly consider the viscosity behavior of POP, POB, and POE homopolymers.

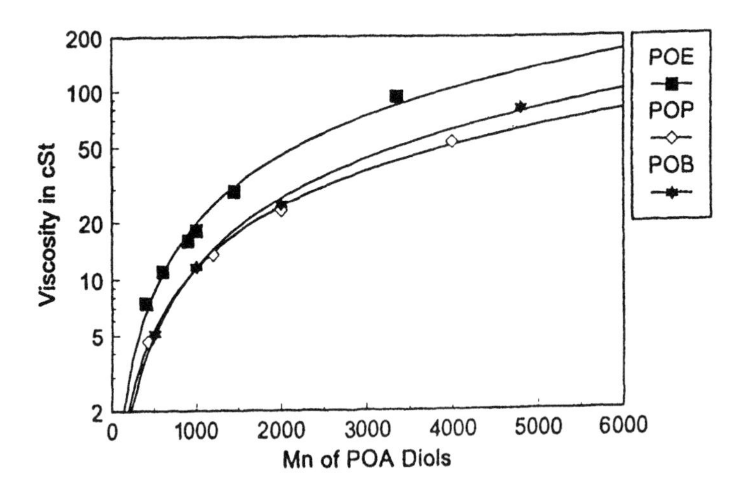

FIG. 5 Kinematic viscosity at 210 °F as a function of average molecular weight for POE, POP, and POB diol homopolymers. *Source*: adapted from [38].

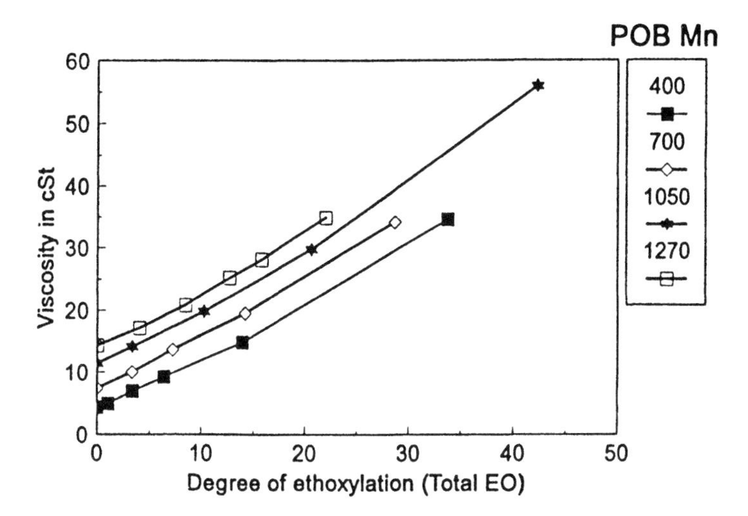

FIG. 6 Kinematic viscosity of E-B-E copolymers at 210 °F as function of the degree of ethoxylation.

Figure 5 shows kinematic viscosity data at 210 °F for homopolymer diols made from EO, PO, and BO [38] as a function of number average molecular weight. Overall, the amorphous POP and POB diols have similarly shaped curves. POE exhibits a marked increase in viscosity over the molecular weight range, presumably due to its highly polar, crystalline nature.

As expected, the viscosity of POP and POB hydrophobes increases with degree of ethoxylation. Figure 6 graphically illustrates this behavior as exemplified by four sets of POB/POE triblock copolymers having 400, 700, 1050, and 1270 mw POB hydrophobes respectively. The same viscosity trend is seen for Pluronic POP/POE triblock liquids, pastes, and solids [39] (Fig. 7). Reverse POB/POE and POP/POE materials behave similarly, with the effect of molecular weight being the overriding factor in viscosity change.

Simek et al. [40] showed that melt viscosity of P-E-P–type polymers can be calculated with reasonable accuracy using POP and POE homopolymer diol viscosity data; the primary source of error was intrinsic friction differences.

The first comprehensive study on aqueous solution viscosity of PEP materials was done by Schmolka and coworkers [41] who studied the effects of concentration and temperature on a variety of Pluronic polyols. Gel formation and viscosity maxima were noted for certain materials. Later studies [42–44] have shed light on rheological changes in gel-forming E-P-E systems and relationships that exist between POE solution viscosity and gel structure. Bloss and coworkers [45] have used aqueous viscometric data in discussing the "core-

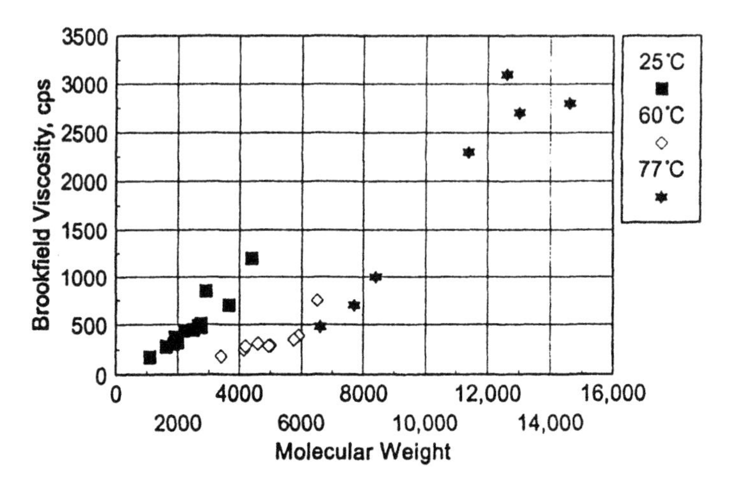

FIG. 7 Brookfield viscosity of Pluronic E-P-E copolymers as a function of total molecular weight. *Source*: adapted from [39].

shell" nature of E-P-E copolymer micelles resulting from POP/POE incompatibility in solution.

C. Polyoxyalkylene Polarity

Nonionic surfactants may be classified in a number of ways, one of which is water solubility. Within a particular class of surfactant, the hydrophobic moiety can be held constant while weight percentage of hydrophilic POE (and therefore POE block molecular weight) is varied. Water solubility, among other properties, changes according to hydrophilic content of the nonionic surfactant.

POP diols having molecular weights between 400 and 500 are fully water soluble. At higher molecular weights, water solubility decreases and hydrocarbon oil solubility tends to increase, but not appreciably. In contrast, low molecular weight POB diols are sparingly water soluble and are miscible with mineral oil and other hydrocarbon solvents in virtually all proportions.

Petrik et al. [46] showed water sorption to be a function of EO content in P-E-P block copolymers. POE homopolymers absorbed the highest amounts of equilibrium moisture followed by P-E-P block copolymers. POP homopolymers absorbed the least amount of water. Houlihan et al. [132] reported on the relative hydrophobicity of several E-P-E block copolymers as measured by pyrene solubilization and carbon black dispersing ability. Semi-empirical equations were derived relating pyrene and carbon black dispersivity to the number of PO and

EO monomer units (hydrophobic and hydrophilic contributions respectively) in the copolymer.

The predictive and useful hydrophile-lipophile balance (HLB) system [47] has been applied to nonionic ethoxylates in very loose terms, and has been questioned as to its practical utility in some situations [48]. A nominal HLB value for ethoxylated hydrophobes can be calculated by dividing the weight percentage of EO by 5. In essence, the calculated HLB is a direct function of EO content. Higher HLB values are therefore indicative of higher water solubility. Compared to simple long-chain fatty alkyl hydrophobes, the higher percentage of oxygen in POP and POB hydrophobes is undoubtedly a factor in the necessity of having higher molecular weights to obtain adequate levels of surfactant performance. Presence of oxygen from the ether units of POA hydrophobes changes solubility characteristics such that higher polarity per unit length is realized. Within in this context it is reasonable to say that the hydrophobic moiety represented by one class of nonionic ethoxylate does not necessarily possess the same solubility characteristics as one found in a different class. For example, a C_{12}–C_{13} fatty-chain hydrophobe would be expected to exhibit greater oil solubility than a polyoxypropylene hydrophobe of similar molecular weight. In fact, the POP "hydrophobe" of this molecular weight range would instead be water soluble. However, when using the simple equation mentioned above to determine the HLB, the resulting values would be the same and the rule of thumb would not be useful.

Various methods have been developed to determine HLB of surfactants experimentally [49–52] by emulsion inversion and dielectric constant, without the necessity of performing exhaustive emulsification experiments [53]. However, data obtained from these methods such as water number and cloud point were eventually correlated with "known" HLB values. Aqueous phenol titration of polyoxyethylenated hydrophobes to a cloudy end-point correlated with POE content [54]. Szymanowski [55] used Lin and Marszall's "hydrophobicity index" values [56] in the study of sulfated and quaternized fatty alcohol ethoxylates.

1. Polarity Index Measurement by Inverse Gas Chromatography

The method of inverse gas chromatography (IGC) has proven to be useful in the study of polymeric systems in general [57] and surfactant polarity in particular. The concept involves the study of retention characteristics of low molecular weight organic probes of varying polarity on a polymeric gas chromatographic stationary phase. The relative retention of polar and nonpolar probes yields information on stationary phase polarity.

Several examples of POA stationary phases have been reported. Early work by Jennings [58] showed that Pluronic F-68 was useful as a liquid phase for gas chromatography (GC) separation of milk flavorings. Bevilacqua [59] studied the retention of several volatile polar compounds using Pluronic P-84, P-66, P-75, L-81, and P-84 stationary phases. Pluronic F-68 was studied by Hlavay et al. [60] and found to be more stable than Carbowax® 20M as a liquid phase for separation of essential oils. Grob and Grob [61] found utility in the use of Pluronic L-61, L-64, and F-68 as liquid phase coatings for capillary GC columns and mentioned polarity increase as a function of weight percent EO. Tenney [62] reported the selectivity of a number of stationary phase liquids for numerous mixtures of polar and nonpolar compounds. Examples of stationary phases included polypropylene glycol diols of 400 and 2000 molecular weight.

Inverse gas chromatography has also been used to elucidate relative polarity values for several nonionic surfactants. These polarity values could then be correlated with other properties or be used to explain various phenomena. Heubner developed a method for determining relative polarity of a variety of compounds (many of which were surfactants) using IGC [63]. His method was simple, straightforward, and useful. Becher and Birkmeier [64] later used a variation of the method to correlate polarity index with known HLB values of fatty alcohol ethoxylates, sorbitan fatty acid esters, and ethoxylated sorbitan fatty acid esters. Wisniewski and Szymanowski [65] used Heubner's approach to study the polarity of alkyl phenyl amine ethoxylates. In related work, Reinhardt and Wachs [66] studied the relationship between IGC retention indices and HLB using water and aliphatic hydrocarbons as probe materials. Fineman [67] used Heubner's approach for determining relative hydrophobicity of several nonionics including fatty alcohol ethoxylates (AE) and nonyl phenol ethoxylates (NPE). A definite difference in hydrophobe polarity was seen between these two classes, with the nonyl phenyl hydrophobe showing the highest polarity. Broniarz et al. [68,69] used the same approach in their study of several nonionic surfactant classes and found, similarly to Fineman, the same order of polarity between AE and NPE. Szymanowski et al. [70] correlated Heubner's polarity indices with surface tension, water value, emulsion inversion point, foaming, and wetting properties of triblock E-B-E copolymers.

Polarity scales derived from testing a host of GC liquid phases, including some polyoxyalkylene homopolymers, have been reported [71,72]. These scales can be useful in determining the best stationary phase for separating a given mixture of analyte materials. However, to develop an absolute polarity scale would be an elusive endeavor, and fortunately unneeded. Also, great care must be taken in any attempt to correlate polarity values obtained when using different experimental approaches.

2. Polarity Indices of PO/EO and BO/EO Block Copolymers

Many papers have appeared on the study of POB/POE block copolymer polarity [73–80]. Some authors focused on the correlation of POB hydrophobe molecular

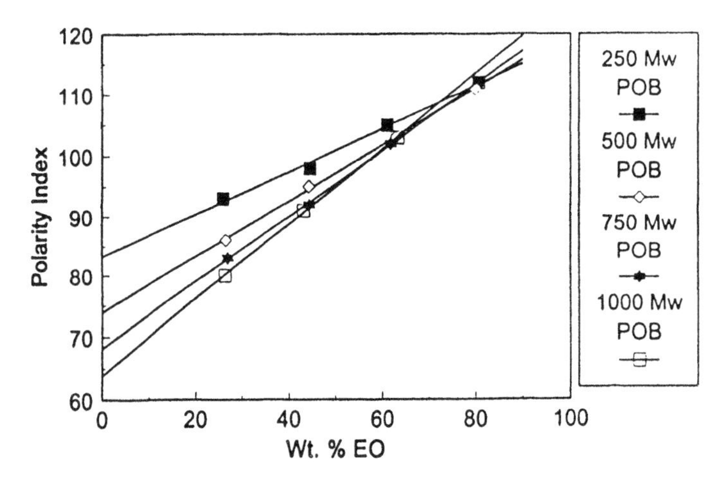

FIG. 8 Polarity index versus weight percent EO for a family of B-E-B block copolymer surfactants. POB molecular weights are per end block. *Source*: adapted from [73].

weight and POE content with polarity index and other properties. Figure 8 shows a plot of polarity index (PI) versus weight percent EO from work done on B-E-B triblock copolymers [73]. As is typical of this type of plot, the curves representing each hydrophobe converge at higher POE content and at a limiting polarity index. Notice that when the curves of Fig. 8 are extrapolated to 0% EO, the lower-limiting PI values of the hydrophobes are obtained, with values ranging from lower to higher depending on the hydrophobe molecular weight (and therefore, the weight percentage of polar hydroxyl groups).

For POA block copolymer nonionic surfactants having a given hydrophobe type, the two primary variables affecting polarity are weight percentage of EO and terminal hydroxyl groups. The dependence of PI on hydrophobe type, percentage of EO, and hydroxyl group content were studied by Nace and Knoell [81]. C_{13}–C_{15} fatty alkyl, nonyl phenyl, POB, and POP hydrophobes along with corresponding ethoxylates were studied by IGC. Heubner polarity indices were plotted against calculated HLB values (HLB = %EO/5) (Fig. 9). Notice that the polarity of the four surfactant classes ranges from POP (highest) to AE (lowest) with the POB hydrophobe falling between POP and nonyl phenyl. The increase in PI for nonyl phenyl at calculated HLB below 5 was attributed to the presence of highly polar residual phenolic OH [81]. It is interesting that the nonyl phenyl hydrophobe has a distinctly higher polarity compared to long-chain alcohol hydrophobes, as other workers have reported [69]. These results show that HLB value alone does not fully describe nonionic surfactants when comparing different classes. However, the polarity distinction between classes becomes less

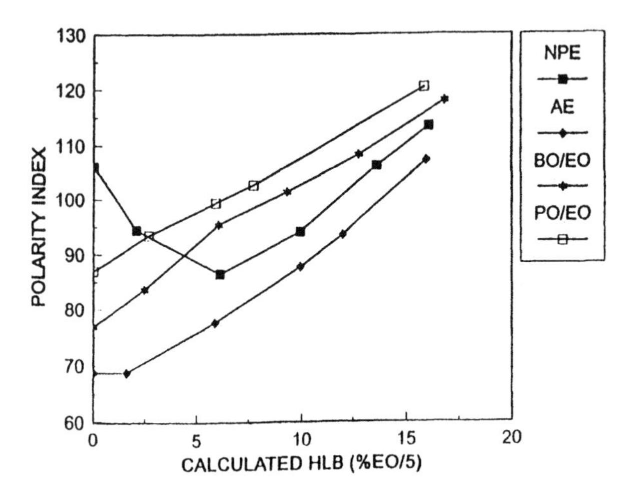

FIG. 9 Polarity index of four classes of nonionic surfactants versus calculated HLB. *Source*: [81] with permission.

clear at high EO levels (Fig. 9), where the curves tend to converge due to dilution of the hydrophobe moiety.

From these observations, it follows that aromatic hydrophobes are somewhat more closely matched to POP and POB hydrophobes in polarity compared to saturated aliphatic hydrophobes. As further evidence of this, Nagarajan et al. [82] found that simple aromatics (e.g., benzene, *o*-xylene, ethyl benzene, and toluene) were solubilized approximately 10 times more efficiently than alkanes (e.g., cyclohexane, hexane, heptane, octane, and decane) using aqueous solutions of POP/POE block copolymers. Also, Szymanowski et al. [136] found that xylene could be emulsified in water more efficiently than mineral oil using E-B-E block copolymer surfactants.

As previously mentioned, terminal hydroxyl groups contribute positively to nonionic surfactant polarity values [81]. Figure 10 shows Heubner polarity indices for POB and POP diols as a function of weight percent OH. As molecular weight of POP and POB homopolymer hydrophobes increased, polarity index decreased and reached limiting lower values. As the relative slopes of the curves in Fig. 10 suggest, hydroxyl group content was found to have more effect on the polarity of POB diols compared to POP diols [81]. It is interesting to note earlier work by Tenney [62] showing that POP 400 diol (8.5% OH) when used as a gas chromatographic stationary phase retained polar probe materials more strongly than POP 4000 (0.85% OH).

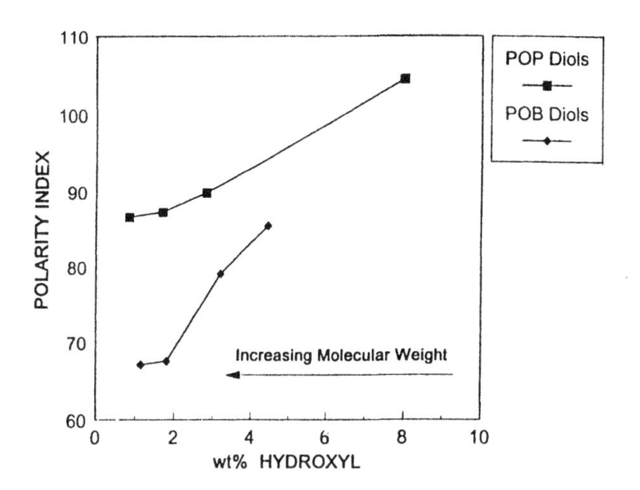

FIG. 10 Polarity index versus hydroxyl content for POB and POP homopolymer diols. *Source*: [81] with permission.

Relative effects of EO and OH content were determined for E-B-E triblock copolymers using the concept of "isopolarity" [81]. They found that a 0.17 unit change in weight percent hydroxyl was approximately equivalent to 1.0 wt% EO. From this it followed that EO content was approximately 6 times more effective than terminal hydroxyl in changing polarity index.

Hydroxyl group effects on various properties have been mentioned in several studies: Interfacial tension between POE and POP homopolymers [83], water solubility of P-E-P block copolymers [84], liquid chromatographic separations of POP/POE copolymers and homopolymers [85], IGC investigations of POE homopolymers [86], micellar solubilization of *n*-octane and *n*-octanol [87], and HLB determination of PO/EO block copolymers [50].

D. Chemical and Thermal Stability

1. Stability in Acid and Caustic Media

In some instances, strongly acidic or basic media can enhance the oxidative degradation of polyethers. The term "stability" can also refer to the solubility of a nonionic surfactant, particularly in systems of high ionic strength. Effects of salt and other ionic species on cloud point and other solution behavior of POP/POE copolymers has been described by several workers [42,88–95]. Other classes of nonionic surfactants [97,98], as well as POE homopolymers [96] can also be affected by salt. Therefore, this phenomenon alone may dictate how

POA nonionics are used in the presence of ionic species. The problem can be handled in some cases by the proper use of other formulants such as coupling agents.

2. Stability to Reducing and Oxidizing Media

Nonionic surfactants having polyoxyethylene hydrophilic groups are somewhat incompatible with oxidizing agents over long periods of time. The stability of POE was reviewed [99] by Donbrow. The generally-accepted mechanism of degradation involves the steps of hydroperoxide formation and subsequent reaction to form aldehyde and carboxylic acid moieties. The presence of trace amounts of acidic and basic species can exacerbate the problem of oxidation. Typical effects of polyether chain oxidation are manifested by the increase of color, and decrease in molecular weight. When neat POA copolymers are handled or stored at high temperatures it is necessary to maintain a nitrogen pad over the material, particularly when acidic or basic species are present.

3. Thermal Stability in Air and Inert Atmosphere

Santacesaria et al. [100] showed by thermogravimetric analysis (TGA) and DSC analysis that stability of polyoxyalkylene surfactants can be related to the number of EO units attached to the hydrophobe. However, no structural data on the PO/EO block copolymers was given except molecular weight.

Data from TGA has been used to compare thermal stabilities of both block copolymers and block copolymer components by observing temperatures at which 10% and 90% weight loss occur. Table 1 lists temperature and weight loss data for three polyoxyalkylene homopolymers and two analogous sets of POP/POE and POB/POE triblock copolymers in the presence of air and nitrogen [115]. In air, the POE diol was somewhat more stable compared to POP and POB diols. POP and POB diols behaved similarly. The increased stability is probably due to presence of a thermal stabilizer. As a class, triblock copolymers were more stable in air than POP and POB homopolymer diols. Within each set of structurally analogous block copolymers, POB/POE diols exhibited slightly increased stability in air. In nitrogen the order of stability for homopolymer diols was POE > > POB > POP. Temperatures at which 10% of the weight loss occurred were higher for POB/POE block copolymers under nitrogen atmosphere. However, 90% weight loss under nitrogen was identical for the lower molecular weight pair of analogs and, in the case of the higher molecular weight block copolymers, the POP/POE diol was the most stable under nitrogen. Overall, as expected, all polymers were more stable under inert atmosphere as compared to under air.

Oxidative stabilization of organic materials is an interesting area, but well beyond the scope of this volume. A few appropriate references on the subject are offered for further reading [100–112].

TABLE 1 TGA Analysis of Homopolymers and Block Copolymers of BO, PO, and EO

			Temperature at specified weight loss			
			Air		Nitrogen	
Copolymer	Wt% OH	Mn	10%	90%	10%	90%
POE 1450	2.34	1450	250	304	453	555
POP 2000	1.70	2000	225	270	313	390
POB 2000	1.70	2000	230	270	356	425
$E_8P_{30}E_8$	1.36	2500	263	315	341	430
$E_8B_{24}E_8$	1.36	2500	270	320	368	430
$E_{34}P_{13}E_{34}$	0.90	3800	247	310	361	475
$E_{34}B_{10}E_{34}$	0.92	3700	273	325	408	455

Rate of temperature rise = 10 °C/min.

III. PERFORMANCE-ORIENTED PROPERTIES

Properties of block copolymers relating to performance issues include wetting, foaming, cloud point, solubilization, emulsification, and phase behavior (which includes gelation). As previously mentioned, a thorough discussion of the subjects of solubilization, emulsification, phase behavior, and cloud point would greatly enlarge this section and fit better in the chapter on physical chemistry. This section will focus on commercially important block copolymer diols and tetrols (normal and reverse block). Diblock monols made from BO, PO, and EO will also be discussed. The reader will find that the spectrum of references given in this section are rather complete and, along with the references cited therein, combine to create an exhaustive bibliography on surfactant properties. References covering block copolymers of other structures, such as random blocks, will not be discussed in light of thorough review given in previous work by Schmolka and Lundsted [113–114]. Data on block copolymers with a functionality of three or more is fairly sparse and, in general, germane to urethane polyol systems. Therefore, these copolymers will not be discussed here, but some information is presented in the chapter on synthesis (Chapter 1).

Representative data will be shown to assist in the formulation of conclusions and principles, such as the comparison of wetting or foaming behavior of two similar surfactant types. Data for certain materials, if not specifically annotated in the text, can be found within cited references.

A. Wetting

In simple terms, wetting is a process of one interface being replaced by another. Spreading, immersional, and adhesional wetting are three types of wetting describable in terms of interface replacement [116]. For example, in the Draves wetting test [117] an air-cotton interface is replaced with a water-cotton interface by immersional means. The presence of a surfactant increases the rate of interfacial replacement. Cotton skeins used in the test are very immune to being wetted by pure water alone, such that the process may take several hours or even days to occur. Adding small amounts of an appropriate surfactant to the water (on the order of 0.1% w/v) can reduce the wetting time to a matter of a few seconds. This standardized test is beneficial for comparing different types of surfactants and also surfactants within the same class which have slight differences in structure, such as degree of ethoxylation. Lower wetting times tend to convey the quality of better performance, provided that experimental conditions are kept constant (hook and skein weight, surfactant concentration, and temperature).

Wetting properties are very important in industrial processes involving textile finishing and particle dispersion. POA block copolymer surfactants can act as quite efficient wetting agents, depending on structure.

1. Wetting Properties of POP/POE Block Copolymers

Wetting properties of PO/EO triblock copolymers of type E-P-E have been described by Schmolka [118,119] in terms of hydrophilic content and molecular mass of the hydrophobe. In general, wetting times decrease (better wetting) as the percentage of hydrophilic EO decreases, and also as the molecular weight of the POP hydrophobe increases. The BASF Corporation uses a graphical method of presenting various properties of Pluronic E-P-E copolymers [35] by plotting POP hydrophobe molecular weight versus weight percentage POE. In the plot for wetting, the region for optimum wetting performance consists of four commercial products (Pluronics L-81, L-92, L-101, and P-103). From data found in the CTFA Cosmetic Ingredients Dictionary [120], the four materials have nominal hydrophobe molecular weights of 2260, 2730, 3130, and 3130, respectively. The CTFA Dictionary uses the names Poloxamer, Meroxapol, and Poloxamine for E-P-E diols, P-E-P diols, and ethylene diamine POP/POE tetrols, respectively. In a table of wetting data [35, p. 25], Pluronic L-81 and L-101 are listed as insoluble (above the cloud point), but assumed to be dispersed adequately enough to act as efficient wetters. Other Pluronic copolymers having notable wetting properties are Pluronic L-62, L-72, P-104, and P-105. Table 2 lists Draves wetting data for Pluronic copolymers at two concentrations [35].

Wetting data reported for ethylene diamine initiated POP/POE block copolymers by Schmolka et al. [118,121] show that Tetronic® polymers with

TABLE 2 Draves Wetting Data for Pluronic Triblock Copolymers

	Draves wetting time (25 °C, sec)	
Pluronic name	1.0%	0.1 %
L-10	6	> 360
L-31	> 360	> 360
L-35	> 360	> 360
F-38	> 360	> 360
L-42	> 360	> 360
L-43	> 360	> 360
L-44	> 360	> 360
L-61	Insoluble	Insoluble
L-62	10	78
L-63	63	> 360
L-64	35	> 360
P-65	> 360	> 360
F-68	> 360	> 360
L-72	1	30
P-75	76	> 360
F-77	> 360	> 360
L-81	Insoluble	Insoluble
P-84	40	260
P-85	36	76
F-87	> 360	> 360
F-88	> 360	> 360
L-92	1	15
F-98	> 360	> 360
L-101	Insoluble	Insoluble
P-103	2	17
P-104	6	30
P-105	16	43
F-108	> 360	> 360
L-121	Insoluble	Insoluble
L-122	12	44
P-123	16	35
F-127	> 360	> 360

Hook and skein weights not specified.
Source: adapted from [35].

TABLE 3 Draves Wetting Data for Tetronic Octablock Copolymers

Tetronic name	Draves wetting time (25 °C, sec) 1.0%	0.1 %
304	> 360	> 360
504	> 360	> 360
701	1	24
702	10	38
704	35	185
707	> 360	> 360
901	1	21
904	8	88
908	> 360	> 360
909	> 360	> 360
1101	1	52
1102	1	15
1104	15	37
1107	> 360	> 360
1301	25	31
1302	4	30
1304	20	48
1307	> 360	> 360
1501	17	181
1502	16	51
1504	39	84
1508	> 360	> 360

Hook and skein weights not specified.
Source: adapted from [35].

hydrophobe molecular weights of approximately 4750 and 20% EO exhibit maximum wetting performance. Table 3 summarizes Draves wetting data for the Tetronic series of copolymers at two concentrations [35, p. 26].

The reverse forms of Pluronic and Tetronic block copolymers (Pluronic R and Tetronic R) result from synthesizing the POE block(s) first, followed by capping with PO to form the final copolymer. In the case of Pluronic R materials, those named as Pluronic 17R1, 25R1, 25R2, 31R1, and 31R2 exhibit the best wetting performance [35, p. 26]. Tetronic 70R2, 90R4, 110R2, 130R2, and 150R4 show maximum wetting for this series [35, p. 27]. Wetting data for Pluronic R and Tetronic R surfactants are shown in Tables 4 and 5.

Kurcharski [122] reported cotton-ring wetting data for hexanol and octanol initiated POP/POE diblock monols. Optimal wetting for hexanol-initiated diblocks was reported for a 15.3 mole PO/12.9 mole EO adduct. Octanol initiated diblocks exhibited better overall wetting. A 6.4 mole PO/7.2 mole EO adduct gave the best wetting results.

Pacifico et al. [123] showed the wetting times (disk method) of several Pluronic materials as a function of concentration and temperature. Wetting times decreased with increasing temperature (30, 60, and 90 °C) for both test concentrations (0.1 and 0.25 wt%). Solubility of the copolymers in water decreased as a function of temperature giving rise to increased hydrophobicity. The net effect was perhaps a better polarity match between the wettable substrate and surfactant, and therefore more efficient wetting.

2. Wetting Properties of POB/POE Block Copolymers

Early data by Spriggs [124] on the wetting properties of POB/POE block copolymers has been discussed in previous work [113,114]. To date only one set of POB/POE block copolymers is commercially available. The B-Series Polyglycol line of products [125] made by the Dow Chemical Company consists of eight products. Six products are water soluble, two being diblock monols of the type B-E and E-B, and four being E-B-E triblock diols. Three product groupings are described by the following nomenclature:

BXX-YYYY specifies a triblock copolymer of the type E-B-E having terminal hydroxyl end groups at each end.

BMXX-YYYY specifies a diblock copolymer of the type E-B in which EO is added to a low molecular weight methyl-terminated alcohol, then capped with BO. In this configuration, a terminal secondary hydroxyl group results.

BLXX-YYYY specifies a diblock copolymer of the type B-E in which BO is added to a low molecular weight butyl terminated alcohol, then capped with EO. This arrangement results in a predominance of primary hydroxyl groups at the polymer terminus.

In this nomenclature

B	=	1,2–Butylene oxide polyglycol hydrophobe
M	=	Methyl-functional initiator
L	=	Butyl-functional initiator
XX	=	Approximate weight percentage of BO in the surfactant
YYYY	=	Approximate molecular weight of the copolymer

TABLE 4 Draves Wetting Data for Reverse Pluronic Copolymers

	Draves wetting time (25 °C, sec)	
Pluronic name	1.0%	0.1%
10R5	> 360	> 360
10R8	> 360	> 360
12R3	> 360	> 360
17R1	30	70
17R2	80	> 360
17R4	> 360	> 360
17R8	> 360	> 360
22R4	> 360	> 360
25R1	3	51
25R2	6	38
25R4	46	100
25R5	> 360	> 360
25R8	> 360	> 360
31R1	1	12
31R2	1	32
31R4	14	62

Hook and skein weights not specified.
Source: adapted from [35].

TABLE 5 Draves Wetting Data for Reverse Tetronic Copolymers

	Draves wetting time (25 °C, sec)	
Tetronic name	1.0%	0.1 %
50R1	35	> 360
50R4	> 360	> 360
50R8	> 360	> 360
70R1	4	168
70R2	38	101
70R4	> 360	> 360
90R1	Insoluble	Insoluble
90R4	10	40
90R8	> 360	> 360
110R1	Insoluble	Insoluble
110R2	3	93
110R7	> 360	> 360
130R1	Insoluble	Insoluble
130R2	1	52
150R1	Insoluble	Insoluble
150R4	16	104
150R8	280	> 360

Hook and skein weights not specified.
Source: adapted from [35].

TABLE 6 Draves Wetting Data for Commercial POB/POE Block Copolymers

Concentration (w/w %)	Draves wetting time (sec)					
	BM45-1600	BL50-1500	B40-1900	B20-3800	B40-2500	B20-5000
0.1	32	41	> 360	> 360	46	> 360
0.5	2.7	7.8	4.8	> 360	9.7	> 360
1.0	1.8	4.5	0	216	6.3	> 360

3 g hook, 5 g skein
Source: adapted from [125].

Table 6 lists Draves wetting data for the copolymers. Note that products having 80% (w/w) EO are least efficient. Diblock monols show the highest overall wetting efficiency.

Pazdzioch et al. [126] and Myszkowski et al. [127] report wetting performance data for POB/POE block copolymers of the type E-B-E, B-E-B, and B-E. Instead of the Draves method, a Polish standardized test was used to measure the concentration needed to sink a standard cotton disk in 100 sec at 20 °C [127]. Therefore, better wetting coincides with lower concentrations. Testing was run above the cloud point for most of compounds. Table 7 lists structural details of the three surfactant classes and corresponding wetting threshold concentrations. Wetting values having an asterisk denote the determination was made below the surfactant cloud point.

Szymanowski et al. [70,128] studied wetting and other properties of E-B-E triblocks. Equations were derived for correlating surfactant performance properties to HLB parameters as measured by IGC, water number, and emulsion inversion point. Conclusions showed that optimum POB molecular weight for wetting was between 900 and 1100, with corresponding POE percentages ranging from 35 to 45. Butanol-initiated B-E copolymers exhibited the best wetting at a hydrophobe molecular weight of approximately 500 and percentage of EO ranging from about 26 to 44. Again, it must be kept in mind that aqueous mixtures of POB/POE copolymers having less than approximately 45–50% EO are typically turbid (above the cloud point) at room temperature.

B. Foaming

Foam is generated when a liquid is agitated in the presence of a gas. As more mixing of the liquid and gaseous phases occurs, more bubbles are created, and

TABLE 7 Draves Wetting Data for E-B-E, B-E-B, and B-E Copolymers

POB block mw[a]	Wt% EO	Total mw	Wetting concentration (g/l) E-B-E	B-E-B	B-E
520	25	700	0.52		0.12
	43	920	0.64		0.16
	62	1360			0.45[b]
	80	2550			0.80
1030	26	1380	0.16		
	44	1820	0.35[b]	0.92	0.21
	61	2650	0.34[b]		0.37[b]
	80	5070	0.68[b]		0.80
1530	26	2060	0.72	0.71	
	44	2720	0.64	0.17	
	62	3990	0.46[b]	0.63	
2030	26	2740			
	44	3620	0.30[b]		
	62	5330		0.63	

[a] Divide by two for B-E-B series.
[b] Test run below the cloud point.
Source: adapted from [126].

therefore more interfacial area. In aqueous systems foam stability is most affected by the presence of surfactants. Foaming behavior can be studied using various static and dynamic methods. The most accepted method is the one-pass, static Ross-Miles standard [129]. In most industrial processes foaming is very undesirable and can lead to low yields, shut-down, or even equipment failure. Foaming is desirable in other applications such as shampoos [119,130] and hand dish-washing liquids. A theoretical background on foaming can be found in several texts, including those of Rosen [116] and Schick [131].

1. Foaming Properties of POP/POE Block Copolymers

The Pluronic and Tetronic lines of POP/POE block copolymers can be considered moderately to low foaming [119]. Moreover, some of the products are useful as foam control agents. Trade literature values [35] give Ross-Miles initial foam height data (t = 0 presumed) at 50 °C and 0.1% concentration. Foam heights range from 0 to 100 mm with maximum foaming shown by Pluronic P-75, F-77, P-84, F-87, and F-88. Pluronic R copolymers are lower foaming

TABLE 8 Ross-Miles Foaming Data for Commercial POB/POE Block Copolymers

Concentration (w/w %)	Temp. (°C)	Ross-Miles foam height, $t = 0/t = 5$ min. (mm)					
		BM45-1600	BL50-1500	B40-1900	B20-3800	B40-2500	B20-5000
0.1	25	116/70	91/33	71/21	96/59	114/45	101/95
1.0	25	194/152	169/33	170/18	162/80	161/44	185/165
0.1	50	106/19	92/15	81/11	144/65	101/13	117/70
1.0	50	191/27	158/22	171/10	187/13	149/16	172/81

Source: adapted from [125].

compared to Pluronic materials, with maximum foam heights of 37 mm. Reverse Pluronics having 10–30% EO exhibit little or no foam [35]. Dynamic foaming data of Pluronic R copolymers at 400 ml/min show a maximum for structures containing approximately 40% EO [118]. Foam heights for the Tetronic series range from 0 to 80 mm. Reverse Tetronic copolymers are virtually nonfoaming except for those designated as Tetronic 90R4 and 150R4, whose values are 20 mm [35]. Schmolka and Seizinger report dynamic foam height data for Tetronic copolymers at flow rates of 400 and 200 mL/min. [121].

2. Foaming Properties of POB/POE Block Copolymers

Schmolka et al. presented a paper on foaming and thickening properties for a series of POB/POE diols (Butronic® polyols from the BASF Corporation) [135]. This line of products has not heretofore been commercialized. Pazdzioch et al. [126] studied the foaming behavior of E-B-E, B-E-B, and B-E copolymers, and found them stable foamers, with B-E copolymers showing the highest foaming as a structural class. Szymanowski et al. [70,128] studied foaming properties of E-B-E materials in relation to surfactant polarity. They showed that for high foaming to occur, the POB hydrophobe should be approximately in the range of 800 to 2000 and the weight percentage of EO in the molecule above 70%. Ross-Miles foaming data for six POB/POE block copolymers is provided in technical literature from The Dow Chemical Company [125] and is reproduced in Table 8 (nomenclature same as above). Foam stability is greatest for BM45–1600 and B20–5000 at 1.0% concentration and 25°C.

Myszkowski et al. [127] found that optimum foaming for *n*-butanol initiated B-E copolymers occurred when hydrophobe molecular weight was between 500 and 1000, and above 62 wt% EO.

C. Comparisons of Copolymers Based on POP and POB Hydrophobes

Data directly comparing POP/POE and POB/POE copolymer properties is beneficial for the application chemist. It is known from IGC and other experiments that POP is more polar than POB [81] and therefore has higher water solubility. It is no surprise that a POB hydrophobe of given molecular mass will be more hydrophobic than a POP molecule of identical molecular mass. For a given hydrophobe molecular mass, it is reasonable to expect increased surface activity from POB/POE versus POP/POE block copolymers.

Nace compared POB/POE block copolymer properties to those of POP/POE copolymers of similar block structure [133]. A commercial literature piece also summarizes these comparisons [134]. Figure 11 shows the structure of four pairs of block copolymers, each showing a POB/POE and analogous POP/POE copolymer having similar hydrophobe molecular weights and weight percentages

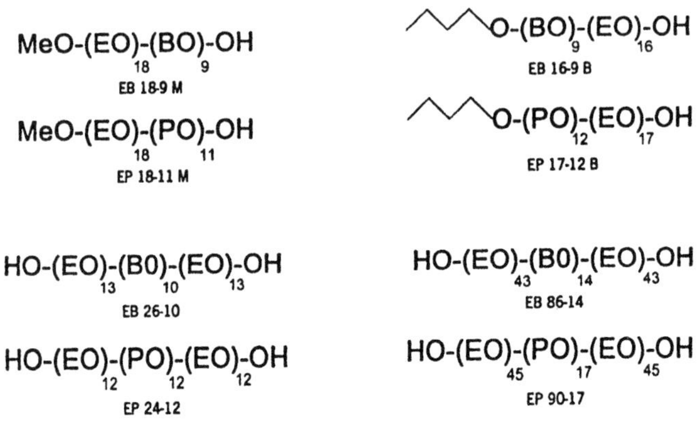

$MeO\text{-}(EO)_{18}\text{-}(BO)_{9}\text{-}OH$

EB 18-9 M

$MeO\text{-}(EO)_{18}\text{-}(PO)_{11}\text{-}OH$

EP 18-11 M

$O\text{-}(BO)_{9}\text{-}(EO)_{16}\text{-}OH$

EB 16-9 B

$O\text{-}(PO)_{12}\text{-}(EO)_{17}\text{-}OH$

EP 17-12 B

$HO\text{-}(EO)_{13}\text{-}(B0)_{10}\text{-}(EO)_{13}\text{-}OH$

EB 26-10

$HO\text{-}(EO)_{12}\text{-}(PO)_{12}\text{-}(EO)_{12}\text{-}OH$

EP 24-12

$HO\text{-}(EO)_{43}\text{-}(B0)_{14}\text{-}(EO)_{43}\text{-}OH$

EB 86-14

$HO\text{-}(EO)_{45}\text{-}(PO)_{17}\text{-}(EO)_{45}\text{-}OH$

EP 90-17

FIG. 11 Structures of analogous POB/POE and POP/POE block copolymer surfactants for property comparison. *Source*: [133].

of EO. Two numbers following the "EB or "EP" prefix refers to the moles of EO and BO (or PO) respectively in the total molecule. The "M" and "B" suffix refers to methyl and *n*-butyl terminations respectively.

Draves wetting data [133] for copolymer analogs is summarized in Table 9. In these comparisons, POB/POE copolymers wetted cotton much better than POP/POE counterparts, except in the case of EB86–14 (80 wt% EO).

TABLE 9 Draves Wetting Performance for POB/POE and POP/POE Block Copolymer Analogs

Copolymer surfactant	Draves wetting time (sec, 1.0%)
EB 18-9 M	1.8
EP 18-11 M	> 360
EB 16-9 B	4.5
EP 17-12 B	> 360
EB 26-10	0
EP 24-12	> 360
EB 86-14	> 360
EP 90-17	> 360

Source: adapted from [133].

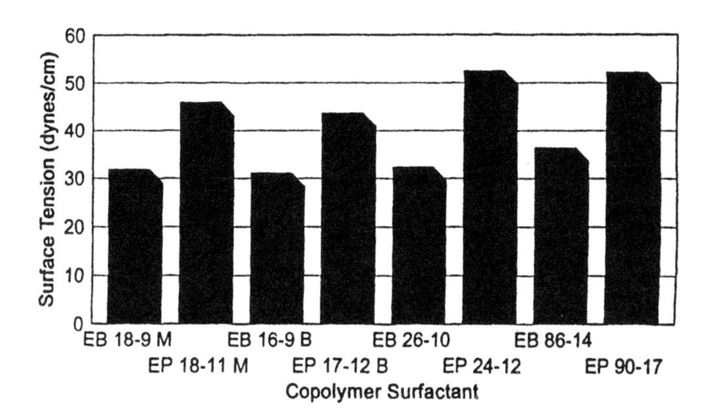

FIG. 12 Aqueous surface tension values for POB/POE and POP/POE copolymer surfactant analogs. *Source*: [133].

Table 10 summarizes data from Ross-Miles foam testing of the four copolymer analog sets [133]. All in all, POP/POE copolymers were lower foaming and showed lower foam stability.

Figures 12 and 13 graphically illustrate the aqueous/air surface tension and aqueous/*n*-dodecane interfacial tension values for POB/POE and POP/POE surfactant analog pairs. Nace reported that surface tension values for POB/POE copolymers were 40–60% lower compared to POP/POE analogs, and interfacial

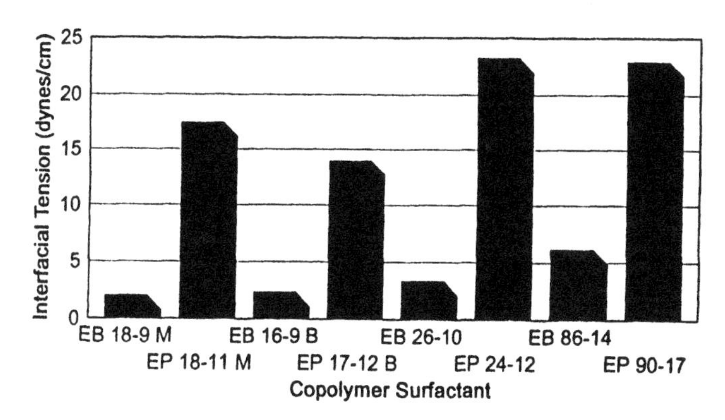

FIG. 13 Interfacial tension values for POB/POE and POP/POE block copolymers against *n*-dodecane. *Source*: [133].

TABLE 10 Ross-Miles Foam Data for POB/POE and POP/POE Block Copolymer Analogs

Copolymer surfactant	Foam height, $t=0, t=5$ (0.1%, mm, 25 °C)	Foam stability $(t=5 / t=0) \times$ 100%, 0.1%, 25 °C	Foam height, $t=0, t=5$ (1.0%, mm, 25 °C)	Foam stability $(t=5 / t=0) \times$ 100%, 1.0%, 25 °C	Foam height, $t=0, t=5$ (0.1%, mm, 50 °C)	Foam stability $(t=5 / t=0) \times$ 100%, 0.1%, 50 °C	Foam height, $t=0, t=5$ (1.0%, mm, 50 °C)	Foam stability $(t=5 / t=0) \times$ 100%, 1.0%, 50 °C
EB 18-9 M	116,70	60	194,152	78	106,19	18	191,27	14
EP 18-11 M	21,0	0	15,0	0	8,0	0	0,0	0
EB 16-9 B	91,33	36	169,33	20	92,15	16	158,22	14
EP 17-12 B	32,12	38	49,19	39	37,9	24	—	—
EB 26-10	71,21	30	170,18	11	81,11	14	171,10	6
EP 24-12	63,4	6	92,6	6	56,8	14	62,4	6
EB 86-14	101,95	94	185,165	89	117,70	60	172,81	47
EP 90-17	82,22	27	84,5	6	60,12	20	64,6	9

Source: adapted from [133].

TABLE 11 Interfacial Performance Data Comparing POB/POE and POP/POE Block Copolymer Surfactants

Copolymer surfactant	cmc (mg/L)	cmc (molar)	Surface excess (moles/cm^2, E-10)	Area per molecule (Å^2)	Δ G mic. (kJ/mole)	cmc/C_{20}	pC_{20}	Maximum surface pressure (dynes/cm)
EB 18-9 M	35	2.3 E-5	2.2	75	-36.4	29.6	6.1	40
EP 18-11 M	90	6.0 E-5	1.2	136	-34.0	7.2	5.1	26
EB 16-9 B	18	1.2 E-5	3.3	51	-37.9	15.1	6.1	41
EP 17-12 B	611	4.1 E-4	0.8	219	-29.3	81.9	5.3	28
EB 26-10	380	2.1 E-4	1.6	103	-30.9	163.7	5.9	40
EP 24-12	899	5.0 E-4	0.5	306	-28.8	0.9	3.3	20
EB 86-14	190	4.0 E-5	1.6	107	-35.0	75.9	6.3	36
EP 90-17	411	8.2 E-5	0.5	323	-33.2	0.4	3.7	20

Source: adapted from [133].

tension was 300–700% lower. The differences were attributed to increased lipophilic character of the POB hydrophobe [133].

Differences in interfacial performance of POB/POE and POP/POE block copolymers were also studied [133]. Table 11 summarizes values of cmc, surface excess concentration, area per molecule at the interface, free energy of micellization, cmc/C_{20}, pC_{20}, and maximum surface pressure. All of these parameters are well-described by Rosen [116]; cmc values were lower for the POB/POE copolymers, as were the free energies of micellization. POP/POE copolymers occupied 2–4 times more area at the interface, perhaps due to a higher degree of hydration. Surfactant efficiency, as manifested by the pC_{20} was 1–2 orders of magnitude higher for POB/POE copolymers. Effectiveness (maximum surface tension lowering) of POB/POE block copolymers was 13–20 dynes/cm higher compared to POP/POE analogs.

Nace also studied performance properties of POP/POE block copolymers (type E-P-E) having POP hydrophobes of 2000–3000 molecular weight [133]. As expected, surfactant properties were improved over lower molecular weight E-P-E materials studied in the direct comparison. However, cmc values were higher compared to POB/POE copolymers. These findings indicate that POP hydrophobes must be in the range of 2000 to 3000 molecular weight (versus 750–1000 for POB) to approach the same level of surfactant performance. It is also known that higher molecular weight ethoxylated POP hydrophobes can have drawbacks such as isomerization side reactions which may give rise to potentially undesirable monofunctional diblock species.

In summary, the following can be offered regarding differences in POB/POE and POP/POE block copolymers:

- enhanced surfactant behavior due to the oil soluble POB hydrophobe
- lower cmc values on a weight basis translates into using less to obtain micellar aggregates
- smaller areas per molecule at the water/air interface (closer interfacial packing)
- better wetting
- lower surface and interfacial tension
- POB/POE copolymers produce higher amounts of foam and exhibit greater foam stability

REFERENCES

1. M. J. Rosen and Z. H. Zhu, *J. Am. Oil Chem. Soc. 70*:65 (1993).
2. C. C. Price, R. Spector, and A. L. Tumolo, *J. Pol. Sci., Part A-1 5*:407 (1967).

3. R. H. Beaumont, B. Clegg, G. Gee, J. B. M. Herbert, D. J. Marks, R. C. Roberts, and D. Sims, *Polymer 7*:401 (1966).
4. F. E. Bailey, Jr. and J. V. Koleske, in *Alkylene Oxides and Their Polymers* (F. E. Bailey, Jr. and J. V. Koleske, Eds.), Marcel Dekker, New York, 1991, pp. 166–170.
5. H. Tadokoro, *Macromol. Rev. 1*:119 (1967).
6. J. P. Arlie, P. Spegt, and A. Skoulios, *Die Makromol. Chemie 104*:212 (1967).
7. P. Spegt, Die Makromol. *Chemie 140*:167 (1970).
8. D. R. Beech, C. Booth, D. V. Dodgson, R. R. Sharpe, and J. R. S. Waring, *Polymer 13*:73 (1972).
9. D. R. Beech, C. Booth, C. J. Pickles, R. R. Sharpe, and J. R. S. Waring, *Polymer 13*:246 (1972).
10. P. C. Ashman and C. Booth, *Polymer 13*:459 (1972).
11. P. J. Flory, *J. Chem. Phys. 17*:223 (1949).
12. P. J. Flory and A. Vrij, *J. Am. Chem. Soc. 85*:3548 (1963).
13. J. N. Hay, M. Sabir, and R. L. T. Steven, *Polymer 10*:187 (1969).
14. D. R. Beech and C. Booth, *Polymer Lett. 8*:731 (1970).
15. L. Simek, S. Petrik, F. Hadobas, and M. Bohdanecky, *Eur. Polym. J. 26*:371 (1990).
16. V. M. Nace, R. H. Whitmarsh, and M. W. Edens, *J. Am. Oil Chem. Soc. 71*:777 (1994).
17. A. M. Afifi-Effat and J. N. Hay, *J. Chem. Soc., Faraday Trans. II 68*:656 (1972).
18. P. C. Ashman and C. Booth, *Polymer 14*:300 (1973).
19. C. Booth, R. C. Domszy, and Y. K. Leung, *Makromol. Chem. 180*:2765 (1979).
20. M. J. Fraser, D. R. Cooper, and C. Booth, *Polymer 18*:852 (1977).
21. A. Marshall, R. C. Domszy, H. H. Teo, R. H. Mobbs, and C. Booth, *Eur. Polym. J. 17*:885 (1981).
22. H. H. Teo, A. Marshall, and C. Booth, *Makromol. Chem. 183*:2265 (1982).
23. C. Booth and C. J. Pickles, *J. Polymer Sci., Polymer Physics Edition 11*:595 (1973).
24. P. J. Flory, B. E. Eichinger, and R. A. Orwoll, *Macromolecules 1*:287 (1968).
25. C. Booth and D. V. Dodgson, *J. Polymer Sci., Polymer Physics Edition 11*:265 (1973).
26. I. Geczy, *Kolor Ert. 24*:276 (1983).
27. C. Booth and C. J. Pickles, *J. Pol. Sci., Polymer Physics Edition 11*:249 (1973).
28. P. C. Ashman, C. Booth, D. R. Cooper, and C. Price, *Polymer 16*:897 (1975).

29. F. Viras, Y. Z. Luo, K. Viras, R. H. Mobbs, T. A. King, and C. Booth, *Makromol. Chem. 189*:459 (1988).
30. U. Leute and T. L. Smith, *Macromolecules 11*:707 (1978).
31. M. Droscher and T. L. Smith, *Macromolecules 15*:442 (1982).
32. Y. Z. Luo, H. H. Teo, and C. Booth, *Makromol. Chem., Rapid Commun. 4*:773 (1983).
33. Y. Z. Luo, R. B. Stubbersfield, and C. Booth, *Eur. Polym. J. 19*:107 (1983).
34. P. C. Ashman and C. Booth, *Polymer 16*:889 (1975).
35. Product Brochure, Pluronic and Tetronic Surfactants, BASF Corporation (1989).
36. J. T. Patton, Jr., U. S. Patent 3,101,374 to Wyandotte Chemicals Corporation (1963).
37. G. C. Berry and T. G. Fox, *Adv. Polymer Sci., 5*:261 (1968).
38. Product Brochure, Industrial Polyglycols and Custom Derivatives Quick Facts, The Dow Chemical Company (1994).
39. Product Brochure, BASF Wyandotte Nonionic Surfactants, BASF Wyandotte Corporation (undated).
40. L Simek, S. Petrik, and M. Bohdanecky, *J. Appl. Polym. Sci. 52*:1905 (1994).
41. I. R. Schmolka and L. R. Bacon, *J. Am. Oil Chem. Soc. 44*:559 (1967).
42. Y. E. Kirsh, V. P. Panov, T. M. Gelfer, V. N. Limarenko, T. A. Dolotova, S. V. Eletskaya, M V. Smirnov, and G. I. Bystirtskii, *Khim. Farm. Zh. 19*:1270 (1985).
43. V. Lenaerts, C. Triqueneaux, M. Quarton, F. Rieg-Falson, and P. Couvreur, *Int. J. Pharm. 39*:121 (1987).
44. M. M. Zgoda, *Acta Polon. Pharm. - Drug Res. 48*:67 (1991).
45. P. Bloss, W. D. Hergeth, C. Wohlfarth, and S. Wartewig, *Makromol. Chem. 193*:957 (1992).
46. S. Petrik, M. Bohdanecky, F. Hadobas, and L. Simek, *J. Appl. Pol. Sci. 42*:1759 (1991).
47. W. C. Griffin, *Soc. Cos. Chemists 5*:249 (1954).
48. A. Graciaa, J. Lachaise, G. Marion, and R. S. Schechter, *Langmuir 5*:1315 (1989).
49. H. L. Greenwald, G. L. Brown, and M. N. Fineman, *Anal. Chem. 28*:1693 (1956).
50. S. Hayashi and T. Fukushima, *Yukagaku 16*:512 (1967).
51. W. G. Gorman and G. D. Hall, *J. Pharm. Sci. 52*:442 (1963).
52. L. Marszall, *Coll. Polym. Sci. 254*:674 (1976).
53. W. C. Griffin, *J. Soc. Cosmet. Chem. 1*:311 (1949).
54. J. V. Karabinos, *Soap Chem. Spec. 31*:50 (1955).
55. J. Szymanowski, *Progr. Coll. Polym. Sci. 63*:96 (1978).

56. I. J. Lin and L. Marszall, *J. Coll. Int. Sci.* *57*:85 (1976).
57. D. R. Lloyd, T. C. Ward, and H. P. Schreiber, Eds., *Inverse Gas Chromatography*, ACS Symposium Series 391, American Chemical Society, Washington, D.C. (1989).
58. W. G. Jennings, *J. Dairy Sci.* *40*:271 (1957).
59. E. M. Bevilacqua, E. S. English, and J. S. Gall, *Anal. Chem.* *34*:861 (1962).
60. J. Hlavay, A. Bartha, G. Vigh, M. Gazdag, and G. Szepesi, *J. Chromatog.* *204*:59 (1981).
61. K. Grob, Jr. and K. Grob, *J. Chromatog.* *140*:257 (1977).
62. H. M. Tenney, *Anal. Chem.* *30*:2 (1958).
63. V. R. Huebner, *Anal. Chem.* *34*:488 (1962).
64. P. Becher and R. L. Birkmeier, *J. Am. Oil Chem. Soc.* *41*:169 (1964).
65. M. Wisniewski and J. Szymanowski, *Colloid Polym. Sci.* *267*:59 (1989).
66. R. Reinhardt and W. Wachs, *Tenside Surf. Deterg.* *5*:125 (1968).
67. I. G. A. Fineman, *J. Am. Oil Chem. Soc.*, *46*:296 (1969).
68. J. Broniarz, M. Wisniewski, and J. Szymanowski, *Tenside Surf. Deterg.* *10*:75 (1973).
69. J. Broniarz, M. Wisniewski, and J. Szymanowski, *Tenside Surf. Deterg.* *11*:27 (1974).
70. J. Szymanowski, E. Pietrzak, K. Prochaska, and B. Atamanczuk, *Tenside Surf. Deterg.* *20*:18 (1983).
71. L. Rohrschneider, *Z. Analyt. Chem., Bd.* *170*:256 (1959).
72. L. Rohrschneider, *J. Chromatog.* *22*:6 (1966).
73. J. Szymanowski, J. Myszkowski, K. Prochaska, and K. Szafraniak, *Tenside Surf. Deterg.* *19*:11 (1982).
74. J. Szymanowski, E. Pietrzak, M. Wisniewski, and K. Prochaska, *Tenside Surf. Deterg.* *20*:188 (1983).
75. J. Myszkowski, J. Szymanowski, K. Szafraniak, W. Goc, and W. Pazdzioch, *Abh. Akad. Wiss. DDR, Abt. Math., Naturwiss., Tech.* *IN*:337 (1987).
76. J. Szymanowski and K. Prochaska, *Fette Seifen Anstr.* *83*:172 (1981).
77. J. Szymanowski, *Fette Seifen Anstr.* *84*:245 (1982).
78. J. Szymanowski, J. Nowicki, and A. Voelkel, *Coll. Polym. Sci.* *257*:494 (1979).
79. J. Szymanowski, J. Myszkowski, A. Voelkel, and W. Pazdzioch, *Chemia Stosowana* *27*:97 (1983).
80. J. Szymanowski and M. Wisniewski, *Fette Seifen Anstr.* *84*:486 (1982).
81. V. M. Nace and J. C. Knoell, *J. Am. Oil Chem. Soc.* *72*:89 (1995).
82. R. Nagarajan, M. Barry, and E. Ruckenstein, *Langmuir* *2*:210 (1986).
83. A. I. Bailey, B. K. Salem, D. J. Walsh, and A. Zeytountsian, *Coll. Polym. Sci.* *257*:948 (1979).

84. L. Simek, S. Petrik, F. Hadobas, and M. Bohdanecky, *Eur. Polym. J. 26*:375 (1990).
85. A. V. Gorshkov, H. Much, H. Becker, H. Pasch, V. V. Evreinov, and S. G. Entelis, *J. Chromatog. 523*:91 (1990).
86. P. Sakellariou, M. H. Abraham, and G. S. Whiting, *Coll. Polym. Sci. 272*:872 (1994).
87. Y. Saito, M. Abe, and T. Sato, *Coll. Polym. Sci. 271*:774 (1993).
88. T. F. Tadros and B. Vincent, *J. Phys. Chem. 84*:1575 (1980).
89. J. Juhasz, V. Lenaerts, P. V. M. Tan, and H. Ong, *J. Coll. Interface Sci. 136*:168 (1990).
90. P. Bahadur, K. Pandya, M. Almgren, P. Li, and P. Stilbs, *Coll. Polym. Sci. 271*:657 (1993).
91. A. Louai, D. Sarazin, G. Pollet, J. Francois, and F. Moreaux, *Polymer 32*:713 (1991).
92. S. Muto, T. Ino, and K. Meguro, *J. Am. Oil. Chem. Soc. 49*:437 (1972).
93. M. Almgren, J. Stam, C. Lindblad, P. Li, P. Stilbs, and P. Bahadur, *J. Phys. Chem. 95*:5677 (1991).
94. P. Bahadur, P. Li, M. Almgren, and W. Brown, *Langmuir 8*:1903 (1992).
95. D. A. Spagnolo and K. T. Chuang, *Can. J. Chem. Eng. 63*:572 (1985).
96. F. E. Bailey, Jr. and R. W. Callard, *J. Appl. Polym Sci. 1*:56 (1959).
97. W. N. Maclay, *J. Coll. Sci. 11*:272 (1956).
98. K. Weckstrom, *Chem. Phys. Lett. 119*:503 (1985).
99. M. Donbrow in *Nonionic Surfactants Physical Chemistry* (M. J. Schick, Ed.), Marcel Dekker, New York, 1987, pp. 1011–1072.
100. E. Santacesaria, D. Gelosa, M. Di Serio, and R. Tesser, *J. Appl. Polym. Sci. 42*:2053 (1991).
101. W. G. Lloyd, *J. Chem. Eng. Data 6*:541 (1961).
102. J. Pospisil and P. Llomchuk, Eds., *Oxidation and Inhibition in Organic Materials,* Volume 1, CRC Press, Boca Raton, FL, 1990.
103. K. Yamaguchi, Y. Ohkatsu, and T. Kusano, *Sekiyu Gakkaishi 34*:458 (1991).
104. P. R. Paolino, Plastics Compounding, *3*:49 (1980).
105. J. D. Dziezak, *Food Tech., Sept.*:94 (1986).
106. C. W. McGary, Jr., *J. Polym. Sci. 46*:51 (1960).
107. D. Al-Sammerrai and N. Nidawy, *Thermochim. Acta 132*:245 (1988).
108. H. Kaczmarek, L. A. Linden, and J. F. Rabek, *Macromol. Symp. 84*:351 (1994).
109. A. M. Afifi-Effat and J. N. Hay, *Eur. Polym. J. 8*:289 (1972).
110. U. Hahner, W. D. Habicher, and K. Schwetlick, *Polym. Degrad. Stab. 34*:111 (1991).
111. U. Hahner, W. D. Habicher, and K. Schwetlick, *Polym. Degrad. Stab. 34*:119 (1991).

112. S. L. Madorsky and S. Straus, *J. Polym. Sci. 36*:183 (1959).
113. I. R. Schmolka, in *Nonionic Surfactants* (M. J. Schick, Ed.), Marcel Dekker, New York, 1967, pp. 300–371.
114. L. G. Lundsted and I. R. Schmolka, in *Block and Graft Copolymerization,* Vol. 2 (R. J. Ceresa, Ed.), John Wiley and Sons, New York, 1976, pp. 1–272.
115. V. M. Nace and J. C. Knoell, unpublished results, The Dow Chemical Company, 1993.
116. M. J. Rosen, in *Surfactants and Interfacial Phenomena,* Second Edition, John Wiley and Sons, New York, 1989.
117. American Society for Testing and Materials, Philadelphia, ASTM Method D2281–68.
118. I. R. Schmolka, *J. Am. Oil Chem. Soc. 54*:110 (1977).
119. I. R. Schmolka, *Am. Perfum. Cosmet. 82*:25 (1967).
120. N. F. Estrin, P. A. Crosley, and C. R. Haynes, Eds., *CTFA Cosmetic Ingredient Dictionary,* Third Edition, The Cosmetic, Toiletry and Fragrance Association, Inc. Washington, D.C., 1982.
121. I. R. Schmolka and R. K. Seizinger, *J. Am. Oil Chem. Soc. 45*:715 (1968).
122. S. Kucharski, *Tenside Surf. Deterg. 11*:101 (1974).
123. C. R. Pacifico and L. G. Lundsted, *Soap Sanit. Chem. 26*:40 (1950).
124. J. S. Spriggs, U.S. Patent 2,828,345 to The Dow Chemical Company (1958).
125. Product Brochure: B-Series Polyglycols, Butylene Oxide/Ethylene Oxide Block Copolymers, The Dow Chemical Company (1992).
126. W. Pazdzioch, J. Myszkowski, K. Szafraniak, and W. Goc, *Prez. Chem. 60*:402 (1981).
127. J. Myszkowski, J. Szymanowski, W. Goc, and K. Alejski, *Tenside Surf. Deterg. 19*:7 (1982).
128. J. Szymanowski, J. Myszkowski, W. Szafraniak, and J. Nowicki, *Tenside Surf. Deterg. 19*:14 (1982).
129. American Society for Testing and Materials, Philadelphia, ASTM Method D1173–53.
130. I. R. Schmolka, *Cosmet. and Toilet. 97*:61 (1982).
131. M. J. Schick, Ed., *Nonionic Surfactants Physical Chemistry,* Marcel Dekker, New York, 1987.
132. P. W. Houlihan, D. Fornasiero, F. Grieser, and T. W. Healy, *Coll. Surf. 69*: 147(1992).
133. V. M. Nace, *J. Am. Oil Chem. Soc. 73*:1 (1996).
134. Product Brochure: Property Differences in BO/EO and PO/EO Block Copolymer Surfactants, The Dow Chemical Company (1994).
135. I. R. Schmolka and B. Kim, Am. Oil Chemists' Soc. Annual Meeting, Chicago, 1983.
136. J. Szymanowski, J. Myszkowski, E. Pietrzak, and K. Prochaska, *Tenside Surf. Deterg. 20*:23 (1983).

5

Applications of Polyoxyalkylene Block Copolymer Surfactants

MICHAEL W. EDENS Industrial Polyglycol Research and Development, The Dow Chemical Company, Freeport, Texas

I. Introduction 186

II. Medical, Biomedical, and Pharmaceutical Applications 186
  A. Gels and drug delivery 186
  B. Controlled drug release 188
  C. Nonfouling surfaces 190
  D. Burn wound dressings 191
  E. Emulsifiers, dispersants, and stabilizers 191
  F. Miscellaneous 192

III. Coal and Petroleum Applications 193

IV. Plastics and Polymers Applications 194

V. Agricultural Applications 195

VI. Emulsion Polymerization Applications 195

VII. Paper and Coatings Applications 196

VIII. Photography Applications 197

IX. Cleaner and Detergent Applications 198

X. Personal Care Applications 199

XI. Metal Cleaning and Corrosion Prevention Applications 200

XII. Miscellaneous Applications 201

References 202

I. INTRODUCTION

The infinite number of combinations of oxide type, oxide ratio, and molecular weight available in polyoxyalkylene block copolymer surfactants gives rise to materials with a wide variety of useful properties. These properties have been utilized by many industries since their introduction in the early 1950's [1–3] by Wyandotte Chemicals Corporation. Application areas for block copolymers range from agricultural chemicals to detergents to food to biomedical and veterinary. This chapter will partially cover the historical applications for block copolymers, but will focus in detail on the most recent applications discussed in the primary and patent literature.

Several excellent books and articles are available covering the historical development of many applications for block copolymer polyol surfactants. Lundsted and Schmolka [4] offer an excellent source of background information on the applications of block copolymer. Other review articles are cited in the specific application sections to follow. The early dominance of the Pluronic® trademark (BASF Corporation) has led to block copolymer polyol surfactants often being referred to in the literature by their Pluronic designation rather than their molecular structure. Another common nomenclature is the generic name poloxamer. In order for the reader to understand and compare products in various applications, Table 1 lists the various Pluronic and poloxamer names for the polyols and gives a physical description of the molecule.

II. MEDICAL, BIOMEDICAL, AND PHARMACEUTICAL APPLICATIONS

A. Gels and Drug Delivery

The thermal gelling properties and low toxicity of block copolymer surfactants [5–7] make them excellent candidates for medical and biomedical applications. Schmolka [8] wrote an excellent review of the use of poloxamers in the pharmaceutical industry with references through 1987, and compared the gels made from ethylene oxide/propylene oxide (EO/PO) and ethylene oxide/1,2-butylene oxide (EO/BO) block copolymers [180]. Henry [9,177] and Leach [46] used polyoxyalkylene gel compositions for reducing postsurgical adhesion formation/reformation in mammals following injury to the organs of the peritoneal, pelvic, or pleural cavity. The gels included an ionic polysaccharide and be formulated to a desired osmolality, with the polyoxyalkylene surfactant not contributing to the osmolality of the solution. The solutions prepared were liquids at room temperature and turned to gels at body temperature. Pluronic F-127 was reported to work well in this application at 10–40 wt% concentration. Block copolymers made from 1,2–butylene oxide and ethylene oxide work in this application when

TABLE 1 Polyoxyalkylene Block Copolymer Nomenclature.

Pluronic[a]	Poloxamer[a]	Hydrophobe mw[b]	%EO[b]
F-68	188	1750	80
F-77	217	2050	70
F-87	237	2250	70
F-88	238	2250	80
F-98	288	2750	80
F-108	338	3250	80
F-127	407	4000	70
L-35	105	5950	50
L-43	123	1200	30
L-44	124	1200	40
L-61	181	1750	10
L-62	182	1750	20
L-63	183	1750	30
L-64	184	1750	40
L-72	212	2050	20
L-81	231	2250	10
L-92	282	2750	20
L-101	331	3250	10
L-121	401	4000	10
L-122	402	4000	20
P-65	185	1750	50
P-85	235	2250	50
P-103	333	3250	30
P-104	334	3250	40
P-105	335	3250	50
P-123	403	4000	30
17R1[c]	171[d]	1410[a]	10[a]
25R2[c]	252[d]	2100[a]	20[a]
25R8[c]	258[d]	2100[a]	80[a]
31R1[c]	311[d]	2450[a]	10[a]

[a] *Source*: [179]
[b] *Source*: [141]
[c] Reverse block copolymers (PPO-PEO-PPO)
[d] Meroxapol

the hydrophobic block has a molecular weight about 3000 and the molecule contains at least 60% EO. Viegas et al. [10–12] used thermoreversable gels to be useful as body cavity drug delivery systems. The gels were especially useful since the gel transition temperature and the rigidity of the gel can be modified by adjustment of the pH as well as the concentration of polymer. A gel solution containing 19% Pluronic F-127 was effective in this application. Joshi et al. [13] investigated the use of Pluronic polyols as gel agents for injectable, droppable, or oral drug delivery systems.

Ophthalmic drug delivery is reported using a solution of 18% Pluronic F-127 [14]. Gels made from Pluronic F-127 are useful in the surgical correction of corneal astigmatism, myopia, and hyperopia [15]. A corneal contact mask was prepared from the gel and this rendered the corneal surface suitable for laser ablation. Kim et al. [16] used Pluronic F-108 as a material for alloplastic keratorefractive surgery. This technique made use of the gelling properties of the water solutions of the F-108.

B. Controlled Drug Release

An area of interest in biomedical research is controlled delivery of intravenous drugs. Illium and Davis [17] studied colloidal systems such as liposomes, microspheres, nanospheres, and emulsions as drug-targeting devices. The problem with these systems is that they are cleared from the body by the reticuloendothelial system and the capillary beds of the lungs. The authors found that treating the particles, in this case polystyrene microspheres, with nonionic emulsifiers led to slow clearance from the body and an altered distribution pattern to organ sites. Poloxamer 338 was used to coat the microspheres which were then injected into rabbits. Liver uptake of the particles was reduced, and blood activity was extended when compared to uncoated particles. Poloxamer 338 was more effective at coating the particles than Poloxamer 188, probably due to the thickness of adsorbed layers. Stoinik et al. [18] investigated the surface modification of poly(beta-malic acid-cobenzyl malate) spheres with poloxamer and poloxamine stabilizers. They found significant differences in the coating thickness, surface hydrophobicity, and zeta potential between the two products. Coombes et al. [19] found poloxamers could be used as surfactants in an emulsification technique to prepare poly(DL-lactide coglycolide) microspheres. This resulted in formation of the particles and a simultaneous coating of the particles with surfactant. Norman et al. [20] reported coating polystyrene microspheres with Poloxamers 235, 237, 238, and 407 reduced the adsorption of human serum albumin by the spheres. Porter et al. [21] studied the effect of coating small colloidal particulates with Poloxamer 407. He found the coating on 60 nm particles was lower compared to 250 nm particles, perhaps due to the larger surface allowing a less crowded arrangement, and lower surface curvature. These authors found the Poloxamer 407 coating led to a specific sequestering of the particles in the bone marrow of rabbits. They proposed a theory predicting the possibility of a site specific coating. Tan et al. [22] investigated the surface modification of nanoparticles coated with Pluronic F-108 and their interactions with blood components in rats. When the particles were coated, the blood circulation half-life was increased from 20 min to 13 h. Park et al. [23] used Pluronic F-108, P-104, and L-101 to prepare degradable polymeric matrices by blending with poly(L-lactic acid). The water content of the films was controlled by mixing different surfactants to obtain the desired hydrophobicity. When used as drug releasing matrices, the blends showed extended protein release and lower initial protein burst. In later work, Park et al. [24] proposed a gel-like structure within the polymer skeleton which extends protein release and minimizes initial protein burst. Kabanov et al. [25] investigated the design of drug delivery systems using "self-assembling supramoleculecular complexes." These systems were micelles in which the molecules of the drug are solubilized in a Pluronic P-85 or F-68 polyol. These "microcontainers" demonstrated drastic increases in drug effect when alphaglycoprotein was incorporated into a haloperidol-containing micelle. Gilbert and Whiteman [26] used Pluronic polyols as capsule matrix fills. These polyols form gels which act as a barrier to diffusion and serve to control the release of incorporated drug molecules. Miyazaki et al. [27] used Pluronic F-127 as a sustained-release vehicle for intraperitoneal administration of antitumor agents. The gelling properties of the F-127 allowed slow release of the agent into the blood resulting in higher doses than injections without F-127. Moghimi et al. [28] used Pluronic polyols to coat therapeutic agents intended for uptake by the lymphatic system. When these nanoparticle agents were coated, it was possible to control the rate of drainage from the subcutaneous injection site into the lymph nodes. Schwarz et al. [29] reported the same type of research using Poloxamer 188 as a steric stabilizer. Quirion and St-Pierre [30] used Pluronic F-127 as a coating on liposomes to reduce their shell-life and slow clearance from the circulatory system. Calvert et al. [31] used Pluronic F-88, F-87, and P-123 hydrogels for selectively extracting hydrophobic solutes and controlling the release of water-insoluble pharmaceuticals.

Luo et al. [32] studied the micellization, gelation, and drug properties of EO/BO block copolymers. Their results demonstrated polyols made from 15 moles of BO and 80 moles of EO gave sustained release of salicylic acid.

Wu and Miller [33] used poloxamers as components of topical drug delivery systems. Solutions made from these materials gel upon contact with the skin and are used to control the rate of active species release. Poloxamers 188, 238, 288, 333, 334, 335, 338, and 407 were investigated. With Poloxamer 407, higher concentrations of copolymer led to slower release rates and higher solubilization capacity for the lipophilic alkyl nicotinates. Increased temperature resulted in faster release rates. The overall mechanism of release was diffusion through water channels in the gel matrix. Guzman et al. [34] used Pluronic F-108 and F-127 as vehicles for subcutaneous drug delivery systems. Studies in rats indicated an inverse relationship between amount of copolymer and speed of release. Marked reduction in release occurred when F-108 concentration was increased from 20 to 25% (w/v). In one study, the use of a F-127 gel resulted in a sustained plateau plasma level of the drug within ten minutes which lasted for eight hours. Ganesan et al. [35] studied the use of Pluronic F-68 and F-127 to solubilize 1,4–dihydropyridines and obtained formulations demonstrating sustained release after oral administration. Final formulations were 16–20% (w/w) polyol.

C. Nonfouling Surfaces

Surfaces resistant to fouling by proteins and other biomaterials are important in many biotechnological and medical applications. These applications include diagnostic assays, drug delivery systems and medical devices. Sheu et al. [36] prepared a nonfouling polymer resistant surface by treating the low density polyethylene (LDPE) substrate with an inert gas discharge and then immobilization of polyoxyethylene (POE) on the surface. The attachment of the POE was enhanced when the substrate was pretreated with Pluronic L-121, L-122, or F-127. Blainey and Marshall [37] used Synperonic® F-108 to treat a polystyrene substrate and make it resistant to marine bacterial cultures. After four hours, the treated substrate displayed a 70–100% reduction in colonization. After five days, there were no colonies present but isolated bacteria were observed; after two months, no effect was seen. Portoles et al. [38] studied the use of Pluronic F-127 as a bacterial abhesive for hydrogel contact lenses. The surfactant was applied to the lens from a 0.5% solution and was found to decrease the adherence to 60% of the control (100% adherence). When a 4% solution of Pluronic F-127 was used, adherence was reduced to 99% of control. Similar results were obtained with Pluronic P-123. The authors concluded Pluronic F-127 was a promising therapeutic agent for the prevention of implant- or contact lens-related infections. Brink et al. [39] investigated the use of polyol nonionic surfactants as a component of biopolymer-repellent solid surface coatings. Solid surfaces were first coated with polyethylene imine and then reacted with alkylene oxide block copolymers to form coatings resistant to biofouling. Amiji and Park [40] used POE/polyoxypropylene (POP)/ POE triblock copolymers as agents for passivating the surface of biomaterials. The authors irradiated glass in the presence of Pluronic F-68 to obtain a monolayer surface coverage. This surface was found to be very efficient in preventing platelet adherence and activation. Ryan [41] reported a method to reduce the adsorption of platelets and red blood cells by hematology instruments by treatment with Pluronic polyols of 5000–15,000 molecular weight (mw). In a similar fashion, Hagiwara et al. [42] reported the improvement in the blood flow channel sections of medical instruments by treatment (wetting) with Pluronic F-68.

D. Burn Wound Dressings

The use of poloxamers as components of burn wound dressings has received considerable attention. Schmolka [43,44] and Henry [45] reported the concept of using Pluronic F-127 gels as dressings for treatment of burns. Characteristics of these dressings included low (or no) toxicity, enhancement of the healing and repair processes, easy application and removal, and a cleansing effect on the wound. Liquid solutions of the polyol can be prepared and poured directly onto the burn wound. As a liquid, the solution will flow into all depths and contours of the wound, then set as a gel when body temperature is reached. The gel will then control water loss, heat loss, and electrolyte loss. Additives, such as antibiotics, can be applied via the gel. The gel can then be removed with cold water and, due to its surfactant nature, will aid in cleaning the wound.

Rodeheaver et al. [47] demonstrated Pluronic F-68 to be effective as a skin wound cleanser in clinical trials involving over 1000 patients. When solutions of Pluronic F-68 were applied to wounds or burns, patients did not complain of pain and inadvertent spillage into the eye did not cause tearing. Pluronic F-68 also does not interfere with the ability of white blood cells to function in their normal manner. Brenden et al. [48] used EO/PO block copolymer surfactants as components of aqueous chelant compositions for the removal of alginate wound dressings. The surfactants were useful at 0.5 to 5.0 wt% and were water soluble (10 g per 100 mL of water). Surfactants such as Pluronic F-68 do not kill tissue like some anionic or amphoteric surfactants and can be left in contact with the wound. Tranner [49] used Poloxamer 188 as a component of a clear gel facial cleanser which also served as a wound cleanser. The polymer was used at 5 wt%.

E. Emulsifiers, Dispersants, and Stabilizers

Ethylene oxide/propylene oxide block copolymer surfactants find wide use in many biomedical applications as emulsifiers and dispersants. Kelm and Dobrozsi [50] used block copolymers for making dispersions of Tebufelone. This drug is essentially insoluble in water, but must be delivered to the body in such a way as to be absorbable by the gastrointestinal tract. Solid dispersions of the drug and one of several Pluronic polyols resulted in rapid solubilization in gastrointestinal fluids. In this application the block copolymer surfactant was used at 30–50% of the dispersion. Jansen et al. [51] used POP-POE block copolymers to stabilize aqueous solutions of antibodies. Luhtala [52] investigated the use of Poloxamer 184 to retard the crystal growth of carbamazepine in aqueous solutions. When used at 0.5% concentration, more than 50% of the carbamazepine crystals were ≤ 5 μm and the average size was 14 μm. In pure water the average crystal size was 80 μm. Reverse block copolymers were also reported as useful. Hunter et al. [53] used Pluronic polyols to help increase antibody formation to bovine serum albumin in mice when injected in an oil-in-water emulsion. Kabanov et al. [54] used Pluronic micelles as "microcontainers" for in vitro delivery of fluorescein into cells. The surfactant enhanced the ability of the target compounds to penetrate into a cell, and once inside the cells, the target compounds were able to act in their desired manner. Sluka et al. [55] used POE-POP block copolymers as surface active agents to make suspensions for immunoassays in which one of the reactants was present as a solid. The surfactant was present at 0.1–1.0% of the total reaction mixture. Suzuki et al. [56] used oxyethylene-oxypropylene block copolymers as stabilizers for bioactive materials, thus allowing them to be filtered. These copolymers inhibited the agglomeration of the active materials and thus allowed them to pass through membrane filters. Varescon et al. [57] used Pluronic F-68 for emulsifying fluorocarbon-based injectable oxygen carriers. The Pluronic polyol acted synergistically with perfluoroalkylated polyhydroxylated surfactants derived from sugars or polyols. By using synergistic blends of surfactants, the emulsion aging characteristic were greatly improved. Aging factors were reduced from $W = 24.8 \times 10^3$ to $W = 0.68 \times 10^3\ \mu m^3$/day. Tsikurina et al. carried out [58] a detailed study of perfluorodecalin emulsions stabilized by Pluronic F-68.

F. Miscellaneous

Recent literature contains many references to alkylene oxide block copolymers in the biomedical area. Hunter and Duncan [59,60] investigated the use of EO-PO block copolymers as part of a composition effective at not only dissolving blood clots but also effective in reestablishing and maintaining blood flow. The polyols had a polyoxypropylene molecular weight of 1750–3500 and contained 50–90% EO. Hymes et al. [61] used Pluronic F-68 dissolved in Ringer's solution as a therapeutic agent in hemorrhagic shock. Animals treated with 0.4% F-68 in Ringer's lactate plus blood had a significantly higher survival than animals treated with blood alone or Ringer's lactate with blood. Hymes et al. [62] also used POE-POP block copolymers to treat thrombosis in a blood vessel. Solutions of the copolymer served as anticoagulants to dissolve blood clots. The polymer was (minimum) 950 molecular weight PO with at least 50 wt% EO.

Hirschman et al. [63] used Pluronic polyols to inhibit the replication of HIV type 1 virus. Three products, Pluronic L-81, L-92, and L-101 were the most inhibitory, with Pluronic L-81 the most active. The authors observed little debris and few dead cells even when significant inhibition of cell replication occurred.

Nemoto et al. [64] used Pluronic F-68 as a surfactant for coating a fluorine containing membrane to make it hydrophilic. The surfactant was mixed with both a hydrophilic polymer and a water soluble polymer, and is retained on the membrane and within the pores. This treatment was carried out without loss of physical strength of the membrane.

Clarke et al. [65] used POP-POE copolymers as stabilizers to greatly increase the yield of recombinant human albumin in the fermentation of genetically engineered yeast. Concentrations were from 0.5 to 10 g per liter.

Wang and Johnston [66] used Pluronic F-127 to enhance the physical stability of proteins during agitation. Physical instability leads to denaturation and loss of biological activity of the protein. Iijima and Nishimura [67] investigated the use of block copolymers to coat enzymes prior to granulating. This coating served to avoid dust and acted as a barrier to water absorption by the enzyme. The coating quickly dissolved in water, thus making the process suitable for production of enzyme-containing granules for use in detergents, medicines, etc. Müller et al. [68] used block copolymers to prepare a water-soluble metallocene-complex composition for use in cancer therapy. Typical metallocene derivatives were zirconocene dichloride, hafnocene dichloride, and molybdenocene dichloride. El Shaboury [69] reported the increased coating of Pluronic F-68 on tablets and capsules containing Frusemide significantly increased the bioavailability of the material in vivo and in vitro. Stetsko and Chang [70] used Pluronic F-68 to make low-solubility drug compositions. The drugs were coated onto sugar or starch beads which were then coated with poloxamer.

III. COAL AND PETROLEUM APPLICATIONS

The use of block copolymer surfactants for demulsification of crude oil and tars has been reported and reviewed [71–74]. Amaravathi and Pandey [75] reviewed the use of demulsifiers made from alkylene oxide block copolymers reacted with phenol-formaldehyde resins, bisphenol A bis(glycidyl ether), or amino compounds. The use of alkylene oxide block copolymers in other energy-related areas has been studied [76].

Das and Hartland [77] demonstrated the use of a Pluronic polyol (3700 mw) to break the emulsion in a crude oil/salt water mixture. Mohammed et al. [78] used EO/PO block copolymers to demulsify oil from the Buchan oil field. Mohammed had success with a 2500 mw polyol containing 34 PO units and 11 EO units, as well as a 3000 mw polyol with 31 PO and 27 EO units.

Cole [79,80] used Pluronic L-61 in a coal dewatering composition. When the Pluronic L-61 was used at 7–30 wt% along with anionic surfactants and a silicone defoamer, a solution was obtained which was applied directly to coal under ambient conditions to remove water from finely divided ores.

Subburaj et al. [81] applied Pluronic F-127 as a substratum for microbial coal solubilization. Traditional microorganism routes to coal solubilization had problems with separation of the coal particles form the solubilized products and mycelia. With Pluronic F-127, the polyol media was made liquid by reducing the temperature to 4 °C and centrifugation was used for the physical separation.

Moriyama et al. [82] used EO/PO block copolymers for making emulsions of super-heavy oils. The polymers had a PO block (1500 mw) and an hydrophile-lipophile balance (HLB) of 15.8. The emulsions had a low viscosity and were easily atomized for combustion.

Gopalkirshnan and Roznowski [83] used Pluronic F-88 in preparing oil and gas well cementing compositions. A combination of Pluronic F-88 and an alkylphenol ethoxylate (such as Triton® X 405) was synergistic and improved fluid loss, free water, and rheological properties of a styrene/butadiene latex based composition. The surfactants were used at very low levels (i.e., $< 1\%$).

Current [84] used alkylene oxide block copolymer surfactants in enhanced oil recovery for petroleum-bearing formations. A synergistic combination of nonionic and ionic surfactants was shown to reduce the viscosity of crude oil emulsions by as much as 50% compared to surfactant-free controls. The surfactants were water soluble copolymers with viscosity of 490–550 cps at 25 °C. Hsieh [85] used EO-PO copolymer surfactants for preparing stable latex emulsions of ethylacrylate/methylacrylic acid/acrylic acid polymers. These polymers were useful for increasing the viscosity of water injected into petroleum bearing formations for enhanced oil recovery. Pluronic polyols with molecular weight from 1,100 to 14,000 were used successfully at 20% (w/w) concentration.

IV. PLASTICS AND POLYMERS APPLICATIONS

The properties of many types of plastics can be improved by incorporation of alkylene oxide block copolymers during their manufacture or processing [86–88]. Canaday et al. [89] used Pluronic polyols as compatibilizers for polyester polyol and polyol blends. The polyester polyols are incompatible with the fluorocarbon blowing agents used to produce polyurethane foams. This limits the opportunities for using these systems in conventional foam production. Pluronic L-44, P-65, and F-77 were shown to stabilize these blends and gave easily manageable viscosity. The compatabilizers were used at 5–20 wt% of the formulations and, in some cases, gave clear mixtures lasting three weeks.

Funke and Starke [90] reported a copolymer of PO and 30% EO lowered the surface tension of LDPE by almost 5 mN/m, the greatest reduction of any surfactant tested. The EO/PO surfactant also resulted in a doubling of the elongation, a 140% increase in impact bending strength, and a 180% improvement in notch impact strength.

Siloxane-oxyalkylene block copolymers have been shown to be effective in the manufacture of polyurethane foams [169]. Snow et al. [91] used Pluronic polyols with silicone surfactant systems to improve the porosity of flexible polyurethane foams. Blends of 1% Pluronic L-101 produced 10–20% increases in foam porosity versus controls.

V. AGRICULTURAL APPLICATIONS

Block copolymer surfactants have been used extensively as emulsifiers and dispersing agents for insecticides, fungicides, and many other agricultural chemicals [92]. Emulsifiable concentrates are a preferred method of applying plant protective agents, but can not be used in many cases due to the nature of the active ingredient and its requirements for solvents (often flammable) to dissolve it. Frisch and Albrecht [93] used block copolymers to emulsify many active ingredients useful for plant protection. Useful concentrations were typically 2–5% of the formulation. Products included HOE S 3510 and HOE S 1816, both of which are EO/PO copolymers manufactured by Hoechst AG. The resulting stable emulsions contained very small droplets, but remained highly flowable and did not tend toward crystal formation. Ong [94] investigated the use of EO/PO block copolymers to emulsify imidazolinone compounds for use in crop protection concentrates. Interestingly, the preferred polymer for this application was an α-butyl-ω-hydroxy-ethylene oxide-propylene oxide copolymer with molecular weight from 2,400 to 3,500 with an HLB of 12. Commercially this compound is sold as Toximulo® 8320 from Stepan Chemical Co.; concentrations used were 4–5 wt%. Policello [95] reported using a blend of organosilicone and EO/PO copolymers as adjuvants in applications for wetting, dispersing, and spreading pesticides. These blends were synergistic and allowed the formation of good dispersions with low surface tensions and good spreading efficiency. Polyols with molecular weights from 500 to 14,000 containing 10–80% EO were used successfully.

Pinter et al. [96] used nonionic surfactants, including block copolymers, to prepare microemulsions for plant protective coatings. A 1500-mw copolymer mediated the formation of microemulsions containing active ingredients such as organic phosphoric and thiophosphoric acid esters. The nonionic surfactant typically was used at 30 wt% of the emulsion. Schmolka [97] used BO/EO block copolymers as surfactants for preparation of seed protective coatings.

VI. EMULSION POLYMERIZATION APPLICATIONS

French [98] reported in 1958 the use of block copolymer surfactants in emulsion polymerization. A patent on a process to prepare polyvinyl acetate emulsions was obtained in 1961 [99]. In recent work, Niessner et al. [100] of BASF used alkylene oxide block copolymers in the reverse suspension polymerization of water-soluble monomers in hydrocarbons. Older processes produced a coarse particle which required longer than desired drying times. By using a block copolymer of 3000 mw, the investigators were able to polymerize acrylamide in aqueous solution and obtain particles with a mean diameter of 4 mm with a *K* value of 203. Ara and Katoh [101] used POE-POP copolymers as an emulsifying agent for polymerization of acrylic monomers used in metal coating. The copolymer gave improved stability in chromic acid when present at greater than 5% (w/w).

Kim and Luckham [102] used Pluronic P-105 and F-127 as stabilizing agents for polystyrene latexes. Faers and Luckham [181] later extended this study to include the effect of surfactant EO content on emulsion stability. Wasyliw et al. [103] used Pluronic L-61 in the preparation of fortified latexes for use in printing inks and coatings. Nonionic surfactants are typically used in this process as costabilizers and codispersants, but Wasyliw found the low HLB surfactants like Pluronic L-61 suppress the formation of grit and particle aggregates allowing the formation of clear, transparent, and glossy films. Also, these surfactants did not require the use of a defoamer and a less low boiling point alcohol was needed to improve printability. Chen et al. [104] prepared polystyrene and poly(methyl methacrylate) (PMMA) latexes using 0.37 wt% Pluronic F-108. Chen et al. [105] also reported on the role of surfactants in composite latex particle morphology. He found Pluronic F-108 gave a system where the PMMA domains were covered by polystyrene. When nonylphenol ethoxylates were used, the encapsulation was reversed. Hergeth et al. [106] used block copolymer surfactants in styrene emulsion polymerization. The products used were Prohalyt® W 17/80 and W 20/80 from Buna. The author was able to determine the emulsifier was located inside the monomer droplets at system temperatures below the cloud point, and was dissolved in the aqueous phase at temperatures above the cloud point. This phenomenon resulted in a bimodal particle size distribution and an unusual temperature dependence on the reaction rate.

VII. PAPER AND COATINGS APPLICATIONS

Block copolymer surfactants have been components of paper coatings [107–109] and inks [110] for a long time. Kondo and Tsubai [111] used EO/PO copolymers as additives for lithographic printing inks. Without the additive, the ink stained and damaged the ink-receptivity sites of the image portions. The polyols had a molecular weight of 1000–8000 and contained 20–95 wt% PO.

Holland et al. [112] used POE-POP block copolymers as dispersants for aqueous, high percent solids, TiO_2 slurries. The nonionic surfactants were effective when blended with polycarboxylates prepared from acrylic homo- or copolymer, or a maleic copolymer. When used alone, the nonionic surfactants had no dispersion ability and, in fact, thickened the solution. The blends ranged from 90:10 (anionic:nonionic) to 50:50, and the level used was less than 0.6 wt%. Polyols used had a 900–4000 mw PO core and contained 10–80 wt% EO. Bruttel and Kvita [113] used EO/PO block copolymers as coating agents for dyes, fluorescent whitening agents, and other agents. The coating allowed the greatly simplified handling of these agents, gave protection from decomposition, and separated them from other components. The copolymer used contained 80% EO and was applied by spraying a 2 wt% solution onto granules. Gunnell et al. [114] used EO-PO-EO and EO-BO-EO as substituents of xanthene dyes. The polyglycol chains were attached to the xanthene and gave the dye improved lightfastness.

Bast and Scholl [115] used EO/PO block copolymers as chemicals useful for deinking of waste paper by means of flotation deinking. The surfactants were used at 1 wt% concentration. Srivatsa et al. [116] also used EO/PO nonionic surfactants as components of a deinking composition.

Nosaka and Ishiguro [117] used polyglycol surfactants as sizing agents for paper. The molecular weight of the polyols was 600–1200 and PO was present at 60–80 wt%. These polyols were used from 1 to 20 parts by weight in the sizing solution.

VIII. PHOTOGRAPHY APPLICATIONS

Block copolymer polyol surfactants have been used in virtually all areas of photography, from developing solutions to emulsions to dry photographic processes [118,119].

Turner and Riley [120] investigated the use of EO-PO block copolymers in a developing fluid for positive-acting, negative-acting, and reversible lithographic printing plates. Traditional fluids rapidly lose their activity and must be replenished, which interrupts processing. Addition of block copolymer surfactants allows the use of 5–30% alkali without causing damage to substrates and images. In this example, Synperonic® T/304 was used at 0.02% by weight. Tsaur and Kam-Ng [121] used EO/PO block copolymer surfactants to help achieve reduced dispersity tabular grain emulsions in photographic film. The authors used Pluronic L-61, L-35, F-108, F-38, and L-10 to achieve the desired effect. In comparative examples with and without L-61, the coefficient of variation of total grains was reduced from 36.0% to 12.4% when surfactant was present. Surfactant concentrations from 1 to 10% based on the weight of silver are preferred. These authors [122] also used reverse Pluronic polyols 31R1 and 17R1, as well as Pluronic L-63 to achieve very low coefficient of variation tabular grain emulsions. Additional references on this topic are found in Bagchi et al. [123], Tsaur and Kam-Ng [124], and Kim et al. [125].

IX. CLEANER AND DETERGENT APPLICATIONS

Block copolymer surfactants have been used in all aspects of cleaning, including hard surface cleaners, laundry detergents, and rinse aids. [126–128]. Bunczk and Burke [129] used EO-PO block copolymers as part of a phosphate-free liquid lavatory cleansing and sanitizing composition. These cleansers utilized iodine as the sanitizing agent. Typically, phosphates were required to stabilize this type of cleanser, but the use of nonionic surfactants at 5% concentration allowed stable solutions without phosphates. Compounds mentioned were Pluronic L-62, F-68, and P-85. Camp et al. [130] used EO-PO copolymers as binders for the preparation of toilet bowl cleaners containing iodophors. Schmolka [131] used EO/BO block copolymers as a component of high foaming iodophors.

Pancheri and Mao [132] investigated high-sudsing liquid detergent compositions containing a "polymeric surfactant" which was defined as an EO-PO block copolymer. These compositions had superior ability to handle grease due to the presence of the block copolymer. The nonionic surfactant was believed to form complexes with the anionic surfactants present, thus minimizing their ability to leave a micelle or interface once formed. Many Pluronic polyols were mentioned and were used at concentrations of 1–2% (w/w). Holland [133] reported that certain nonionic surfactants used at low levels (2–6% w/w) acted as cosurfactants in cleaners to significantly enhance both oily and particulate soil removal. This synergy allowed replacement of phosphates as detergency boosters in cleaners. Pluronic L-64, P-123, F-108, and F-68 were suitable for this application. Secemski and Lynn [134] reported the combination of an EO-PO block copolymer and a polycarboxylate builder as giving surprising efficiency in the removal of clay soils by nonphosphorous detergents. The polyol was 70% EO and could be a reverse block, such as Pluronic 25R8.

Carmello and Miller [135] used Pluronic L-92 as a coating for caustic beads used as delayed action drain cleaners. The Pluronic L-92 provided a readily water dispersible coating and, when followed by a second, outer coating of hydrophobic material (e.g., talc or calcium stearate), produced a delay in the exothermic reaction which occurred when the caustic beads came into contact with water.

Burke and Roelofs [136] reported a hard surface cleaner formulation using EO-PO copolymer surfactants. The compounds of this invention had an average molecular weight of 3000–7000 and a hydrophile content of 10–60%. These cleaners were especially effective at removing oily soils from glossy or transparent surfaces and leaving the surface streak free without the use of high amounts of solvents or alcohols.

Chang [137] used POE-POP block copolymers such as Pluronic F-127 as processing aids for the manufacture of cleansing blocks for porcelain surfaces such as toilet bowls. The aid was present in the formulation at 2–20%. Gresser

[138] used Pluronic L-64 as an antiredeposition agent and stain removing agent in a detergent composition. Winston et al. [139] used block copolymers as surfactants with foam reducing properties for powdered flux removing compositions. Particularly useful were the reverse blocks, such as Pluronic 25R2, at 0.5 wt% concentration.

X. PERSONAL CARE APPLICATIONS

The physical properties of block copolymer surfactants make them natural candidates for wide use in personal care products. Some of the first commercial uses of EO/PO block copolymers were in the cosmetic industry [140]. While working at BASF Wyandotte, Schmolka obtained over a dozen U.S. patents on the use of EO/PO and EO/BO block copolymers in personal care applications, and has written several review articles on the subject [141–143].

Mori and Makino [144,145] reported a dentifrice composition containing a bactericide and an EO-PO block copolymer surfactant. Typically, to obtain good foam properties in a dentifrice, an anionic surfactant is used. However, in certain cases the anionic surfactant will reacts with the bactericide and causes rapid loss of activity. Pluronic F-88, P-85, and F-87 were used at 25–30% levels giving stable compositions with good foam properties. The gelling properties of the surfactants offered excellent shape retention such that no binder was required. In compositions with an organosilicone-type immobilized bactericide, the surfactants helped emulsify the active ingredient in the aqueous composition and improved the microorganism elimination effects of the bactericide. Nathoo et al. [146] used Pluronic F-127 at 15% concentration to thicken aqueous abrasive oral compositions. Burke et al. [147] investigated the use of Pluronic F-108 and F-127 as foam enhancers for oral compositions which were substantially non-irritating to oral tissue. Their research was prompted by a desire to replace sodium lauryl sulfate as the surfactant in a dental composition in order to make it nonirritating; purified sodium lauryl sulfoacetate was an acceptable surfactant for this application, but low concentrations (0.1–0.3% w/w) of Pluronic F-108 and F-127 were needed to enhance the foam and make the composition acceptable to the consumer. Bianchi et al. [148] used 15–25 wt% block copolymer surfactant as gelling agents for novel fluoride containing dentifrices. Copolymers used were Pluronic F-88, F-98, F-108, and F-127. Nabi et al. [149] used Pluronic F-127 at 0.25 wt% in oral compositions containing salicyanilide antibacterial agents. These agents are water insoluble and incompatible with most oral compositions. The use of block copolymer surfactants along with other surfactants was necessary to prevent phase separation of the salicyanilide. Mitchell and Durga [150] used block copolymer surfactants to stabilize a packaged dental cream. With the advent of polyolefin packages for dental cream, it was necessary to prevent the syneresis that can occur upon contact between the cream and the polyolefin. At 0.5–3% concentration, the copolymer prevented phase separation of the dental cream components. Copolymers preferred for this application were Pluronic F-108, F-87, and L-72.

Several mouthwash compositions have been prepared using EO-PO block copolymer surfactants. Gaffar et al. [151] used Pluronic materials to enhance the prophylactic action of a antibacterial, antiplaque oral mouthwash. Carlin et al. [152] used many Pluronic surfactants in a mouthwash composition containing a bis-biguanido hexane compound as an antibacterial agent. Only specific types of surfactants were useful with these compounds, and some type of surfactant was necessary for proper action of the antibacterial agent. Concentrations of 0.7% (w/v) were typically used. Konopa [153] reported the preparation of a nonalcoholic mouthwash with a homogeneous, uniform appearance and a high degree of bactericidal activity. Many surfactants inhibited the activity of the antimicrobial agent, but Pluronic F-108 at 0.6–1.0% concentration emulsified the active ingredient and did not adversely affect activity. Kleber and Putt [154] used Pluronic F-127 to make anticariogenic oral compositions by employing aluminum in a compatible emulsion system.

Cardin et al. [155] used Pluronic F-68 at 5% concentration in an anti-dandruff shampoo containing a pyridinethione metal salt as the active ingredient. Walele et al. [156] used benzoate esters of polyalkoxylated block copolymers for skin and hair care compositions. Andrews and Kure [157] studied the use of Pluronic F-68 as a cleansing agent in a medicated shampoo formulation for veterinary use. This formulation gave a silky and shining coat on dogs and cats when F-68 was used at 10 wt%.

Piechota [158] used Pluronic F-127 and water for the topical application of water-dispersible active ingredients to skin. A 10–20% solution of F-127 in water gelled in seconds at 80 °F. Active ingredients were put into water and the liquid applied to the skin prior to gel formation.

XI. METAL CLEANING AND CORROSION PREVENTION APPLICATIONS

Metal cleaning has been a traditional area for the use of block copolymer polyol surfactants. Typical uses include cleaning aids and viscosity modifiers for cleaning solutions [159]. Deck et al. [160] reported a low-foaming alkaline cleaner comprising an EO-PO-EO block copolymer and a reverse PO-EO-PO block copolymer. Pluronic L-43 and 31R1 were used at 0.5 to 3 wt% concentration in formulations useful for metal cleaning. The Pluronic L-43 was effective for removing oils from the metal while the reverse block Pluronic 31R1 acted as a defoamer for the cleaning bath. In addition, the reverse block simplified the separation of waste oils from the aqueous treatment solution during effluent treatment.

Schaffhausen [161] used block copolymers esterified with long-chain fatty acids as rust inhibitors in internal combustion engines. Unmodified copolymers also functioned in this capacity, but lacked solubility in crankcase lubricating oils. Preferred polyols were Pluronic L-61 and L-81 with acids such as oleic, linoleic, etc. Mika [162] investigated the use of Pluronic F-127 and L-63 in a solution prepared for removing lead/tin coatings applied to copper or nickel surfaces. The solution used nitric acid to remove the lead/tin from printed wiring boards. The use of the surfactant surprisingly prevented the attack of the nitric acid on the copper or nickel surface. The surfactants were present in the solution at 0.03 g/L.

Ikeda and Kataoka [163] used POE-POP block copolymers as surfactants at 0.1% (w/w) for rust removing formulations on stainless steel surfaces.

XII. MISCELLANEOUS APPLICATIONS

Since their introduction in 1950, block copolymer polyol surfactants have been used in a wide variety of diverse applications [4]. This diversity continues today with many new opportunities for these compounds.

Dahanayake [164] reported a combination of block copolymers and hydrophobic silica to form low-viscosity defoaming/antifoaming agents. Sasaki and Ercillo [165] used block copolymer surfactants to improve the cutability of elastomeric pressure sensitive adhesives. When these adhesives are applied to paper articles and then processed into usable articles, the cutting tools often become coated with adhesive and impair the manufacturing process. When surfactants such as Pluronic F-108 were added at 1–6%, the process was improved. Blackstone et al. [166,178] investigated the use of block copolymer surface active agents to enhance the cooking of wood chips for pulp production. Solutions containing 10% Pluronic L-62 and 7.5% Pluronic F-108 reduced the number of rejects in the pulp by 5%. Mitchell et al. [167] reported similar findings in a 1961 patent. Faulks [168] reported a solution to aid the removal of asbestos from old pipes and other equipment. The solution contained 1% Breox® BL 19–10 as a wetting agent; Breox BL 19–10 is a 1800–2000 mw block copolymer with 8–12% EO. The solution wet the asbestos and allowed removal with only 25% of the fines released when compared to removal without the solution.

Mente [170,171] used reverse block copolymers as thickeners for aqueous liquids, particularly pesticides and paints. He found the reverse blocks served as associative thickeners in these applications, whereas the normal block copolymers did not thicken but actually caused a thinning of the solutions. Compounds found especially effective in this application had a hydrophilic core of 350 mw EO and capped with PO to a total molecular weight of 3550.

DuPont researchers [172] used Pluronic F-108 to wet synthetic pulps. The polyalkylene oxide block copolymer worked alone and in synergistic fashion with silicone surfactants to reduce wetting times from 5 min for untreated pulp to 2 sec. Sink time was reduced from 6 min to 4 sec. Haigh and Harrowfield [173] used block copolymers as lubricants for carding, gilling, combing, and spinning wool. The coefficient of fiber to metal friction was reduced 15% when the lubricant was used versus no lubricant.

Hurter and Hatton [174] used block copolymers as solubilization agents for polycyclic aromatic hydrocarbons. When the hydrophobic/hydrophilic blocks in the surfactant were properly manipulated, the micelle-water partition coefficients were increased. Properly prepared surfactants were used for removal of hydrocarbon contaminants from aqueous waste streams. Ellis et al. [175] studied the use of block copolymers as agents for in situ bioremediation of a creosote contaminated site. Klinger and Milewski [176] used a 1700 mw copolymer with 20% EO as a compatabilizer for anionic and cationic surfactants. Without the nonionic surfactant, the pour points of the cationic/anionic solutions were quite high. The nonionic surfactant displayed a large synergistic effect which resulted in improved foaming characteristics for the solutions.

REFERENCES

1. T. H. Vaughn, H. R. Suter, L. G. Lundsted, and M. K. Kramer, Am. Oil Chemists' Soc. Meeting, San Francisco, September 26, 1950.
2. T. H. Vaughn, H. R. Suter, L. G. Lundsted, and M. K. Kramer, *J. Am. Oil Chemists' Soc. 28*:294 (1951).
3. D. R. Jackson and L. G. Lundsted, U.S. Patent 2,677,700 to Wyandotte Chemicals Corporation (1954).
4. L. G. Lundsted and I. R. Schmolka, in *Block and Graft Copolymerization*, Vol. 2 (R. J. Ceresa, Ed.), John Wiley & Sons, Ltd., London, 1976, pp. 113–272.
5. I. R. Schmolka, *Cosmetics and Toiletries 99*:69 (1984).
6. I. R. Schmolka, U.S. Patent 3,867,533 to BASF Wyandotte Corporation (1975).
7. I. R. Schmolka, U.S. Patent 3,740,421 to BASF Wyandotte Corporation (1973).
8. I. R. Schmolka, in *Polymers for Controlled Drug Delivery* (P. Tarcha, Ed.), CRC, Boca Raton, FL, 1991, pp. 189–214.
9. R. L. Henry, U.S. Patent 5,126,141 to Mediventures Incorporated (1992).
10. T. X. Viegas, L. E. Reeve, and R. L. Henry, U.S. Patent 5,143,731 to Mediventures Incorporated (1992).

11. T. X. Viegas, L. E. Reeve, and R. S. Levinson, U.S. Patent 5,306,501 to Mediventures, Inc. (1994).
12. T. X. Viegas, L. E. Reeve, and R. L. Henry, U.S. Patent 5,346,703 to Mediventures, Inc. (1994).
13. A. Joshi, S. Ding, and K. J. Himmelstein, U.S. Patent 5,252,318 to Allergan, Inc. (1993).
14. T. X. Viegas, L. E.Reeve, and R. S. Levinson, U.S. Patent 5,300,295 to Mediventures, Inc. (1994).
15. T. X. Viegas, L. E. Reeve, and R. L. Henry, U.S. Patent 5,277,911 to Mediventures Inc. (1994).
16. J. P. Kim, R. L. Peiffer, and R. E. Holman, *J. Cataract Refract. Surg. 14*:312 (1988).
17. L. Ilium and S. S. Davis, *J. Pharmaceut. Sci. 72*:1086 (1983).
18. S. Stoinik, M. C. Davies, L. Ilium, S. S. Davis, M. Boustta, and M. Vert, *J. Control. Rel. 30*:57 (1994).
19. A. G. A. Coombes, P. D. Scholes, M. C. Davies, L. Illum, and S. S. Davis, *Biomaterials 15*:673 (1994).
20. M. E. Norman, P. Williams, and L. Illum, *Biomaterials 13*:841 (1992).
21. C. H. Porter, S. M. Moghimi, L. Ilium, and S. S. Davis, *FEBS Lett. 305*:62 (1992).
22. J. S. Tan, D. E. Butterfield, C. L. Voycheck, K. D. Caldwell, and J. T. Li, *Biomaterials 14*:823 (1993).
23. T. G. Park, S. Cohen, R. Langer, *Macromolecules 25*:116 (1992).
24. T. G. Park, S. Cohen, and R. S. Langer, U.S. Patent 5,330,768 to Massachusetts Institute of Technology (1994).
25. A. V. Kabanov, E. V. Batrakova, N. S. Melik-Nubarov, N. A. Fedoseev, T. Y. Dorodnich, V. Y. Alakhov, V. P. Chekhonin, I. R. Nazarova, and V. A. Kabanov, *J. Control. Rel. 22*:141 (1992).
26. J. C. Gilbert and M. Whiteman, in *Proceedings of the International Symposium on Controlled Release Bioactive Materials*, 18 (I. Kellaway, Ed.), Controlled Release Soc., Deerfield, IL, 1991, pp. 574–575.
27. S. Miyazaki, Y. Ohkawa, M. Takada, and D. Attwood, *Chem. Pharm. Bull. 40*:2224 (1992).
28. S. M. Moghimi, A. E. Hawley, N. M. Christy, T. Gray, L. Illum, and S. S. Davis, *FEBS Lett. 344*:25 (1994).
29. C. Schwarz, W. Mehnert, J. S. Lucks, and R. H. Müller, *J. Control. Rel. 30*:83 (1994).
30. F. Quirion and S. St-Pierre, *Biophys. Chem 40*:129 (1991).
31. T. L. Calvert, R. J. Phillips, and S. R. Dungan, *AIChE J. 40*:1449 (1994).
32. Y. Luo, C. V. Nicholas, D. Attwood, J. H. Collett, C. Price, C. Booth, B. Chu, and Z. K. Zhou, *J. Chem. Soc., Faraday Trans. 89*:539 (1993).
33. H. L. Wu and S. C. Miller, *Int. J. Pharm. 66*:213 (1990).

34. M. Guzman, F. F. Garcia, J. Molpeceres, and M. R. Aberturas, *Int. J. Pharm. 80*:119 (1992).
35. M. G. Ganesan, N. R. Desai, G. A. Maier, and P. S. Kulkarni, U.S. Patent 5,160,734 to American Cyanamid Company (1992).
36. M. S. Sheu, A. S. Hoffman, B. D. Ratner, J. Feijen, and J. J. Harris, *J. Adhes. Sci. Technol. 7*:1065 (1993).
37. B. L. Blainey and K. C. Marshall, *Biofouling 4*:309 (1991).
38. M. Portoles, M. F. Refojo, and F. L. Leong, *J. Biomed. Mater. Res. 28*:303 (1994)
39. C. Brink, E. Osterberg, and K. Holmberg, U.S. Patent 5,240,994 to Berol Nobel AB (1993).
40. M. Amiji and K. Park, *J. Colloid Interface Sci. 155*:251 (1993).
41. W. L. Ryan, U.S. Patent 5,250,438 to Streck Laboratories, Inc. (1993).
42. K. Hagiwara, H. Kitoh, Y. Oshibe, and H. Ohmura, U.S. Patent 5,211,913 to Terumo Kabushiki Kaisha (1993).
43. I. R. Schmolka, *J. Biomed. Mater. Res. 6*:571 (1972).
44. I. R. Schmolka, U.S. Patent 4,376,764 to BASF Wyandotte Corporation (1983).
45. R. L. Henry and I. R. Schmolka, *Crit. Rev. Biocompat. 5*:207 (1989).
46. R. E. Leach and R. L. Henry, *Am. J. Obstet. Gynecol. 162*:1317 (1990).
47. G. T. Rodeheaver, L. Kurtz, B. J. Kircher, and R. F. Edlich, *Ann. Emerg. Med. 9*:572 (1980).
48. R. A. Brenden, J. Burkey, and F. T. Kirchner, U.S. Patent 5,292,525 to Merck & Co., Inc. (1994).
49. F. Tranner, U.S. Patent 5,030,374 to International Research and Development Corporation (1991).
50. G. R. Kelm and D. J. Dobrozsi, U.S. Patent 5,281,420 to The Proctor & Gamble Company (1994).
51. T. Jansen, E. J. Janssen, and L. Cornelius, U.S. Patent 4,902,500 to Akzo N. V. (1990).
52. S. Luhtala, *Acta Pharm. Nord. 4*:271 (1992).
53. R. Hunter, F. Strickland, and F. Kézdy, *J. Immun. 127*:1244 (1981).
54. A. V. Kabanov, V. I. Slepnev, L. E. Kuznetsova, E. V. Barrakova, V. Y. Alakhov, N. S. Melik-Nubarov, P. G. Sveshnikov, and V. A. Kabanov, *Biochem. Int. 26*:1035 (1992).
55. P. Sluka, C. Klein, H. Griesser, and U. Kobold, U.S. Patent 5,248,620 to Boehringer Mannheim GmbH (1993).
56. T. Suzuki, H. Ikeda, K. Ikeda, T. Tomono, S. Sekiguchi, T. Ohtani, and S. Suzuki, U.S. Patent 5,219,999 to Mitsubishi Rayon Co. Ltd.; The Japanese Red Cross Society (1993).
57. C. Varescon, A. Manfredi, M. Le Blanc, and J. G. Riess, *J. Colloid Interface Sci. 137*:373 (1990).

58. N. N. Tsikurina, N. M. Zadymova, and Z. N. Markina, in *Proceedings of the Conference on Colloid Chemistry: Memorial to Ervin Wolfram* (E. Kiss and J. Pinter, Eds.), Lorand Eotvos Univ., Budapest, Hung., 1990, pp. 110–113.
59. R. L. Hunter and A. Duncan, U.S. Patent 5,078,995 to Emory University (1992).
60. R. L. Hunter, U.S. Patent 4,897,263 to Emory University (1990).
61. A. C. Hymes, M. H. Safavian, and T. Gunther, *J. Surg. Res. 11*:191 (1971).
62. A. C. Hymes, R. R. Margolis, and R. M. Nalbandiam, U.S. Patent 3,641,240 to Wyandotte Chemicals Corporation (1972).
63. S. Z. Hirschman, M. L. Zucker, and E. Garfinkel, *J. Med. Virol. 42*:249 (1994).
64. Y. Nemoto, S. Yokomachi, and H. Kito, U.S. Patent 5,120,440 to Terumo Kabushiki Kaisha (1992).
65. P. M. Clarke, D. J. Mead, and S. H. Collins, U.S. Patent 5,260,202 to Delta Biotechnology Limited (1993).
66. P. Wang and T. P. Johnston, *J. Parenter. Sci. Technol. 47*:183 (1993).
67. H. Iijima and K. Nishimura, U.S. Patent 4,940,665 to Showa Denko K. K. (1990).
68. B. W. Müller, R. Müller, S. Lucks, and W. Mohr, U.S. Patent 5,296,237 to medac Gesellschaft fur klinische Spzeialpraparate mbH (1994).
69. M. H. El Shaboury, *Acta Pharm. Fenn. 98*:253 (1989).
70. G. Stetsko and K. Chang, U.S. Patent 5,223,268 to Sterling Drug, Inc. (1993).
71. G. P. Canevari and R. J. Fiocco, U.S. Patent 3,331,765 to Esso Research and Engineering Company (1967).
72. H-F. Fink, G. Koerner, and G. Rossmy, U.S. Patent 4,029,596 to Th. Goldschmidt AG (1977).
73. F. Staiss, R. Böhm, and R. Kupfer, *Fossil Fuels, Derivatives, and Related Products 6*:334 (1991).
74. S. E. Taylor, *Chem. Ind. (Lond.), 20*:770–3 (1992).
75. M. Amaravathi and B. P. Pandey, *Res. Ind. 36*:198 (1991).
76. J. H. Niu and I. R. Schmolka, U.S. Patent 4,637,822 to BASF Corporation (1987).
77. P. K. Das and S. Hartland, *Chem. Eng. Commun. 92*:169 (1990).
78. R. A. Mohammed, A. I. Bailey, P. F. Luckham, and S. E. Taylor, *Colloids Surf., A 83*:261 (1994).
79. R. Cole, U.S. Patent 5,048,199 to Wen-Don Corporation (1991).
80. R. Cole, U.S. Patent 4,985,162 to Wen-Don Corporation (1991).
81. K. Subburaj, S. Jayaraman, and M. Gunasekaran, *J. Ind. Microbiol. 9*:225 (1992).

82. N. Moriyama, T. Ogura, and A. Hirake, U.S. Patent 5,024,676 to Kao Corporation (1991).
83. S. Gopalkirshnan and M. Roznowski, U.S. Patent 5,258,072 to BASF Corporation (1993).
84. S. P. Current, U.S. Patent 5,110,487 to Chevron Corporation (1992).
85. W. Hsieh, U.S. Patent 4,970,251 to Union Oil Company of California (1990).
86. W. E. Walles, U.S. Patent 2,832,696 to The Dow Chemical Company (1958).
87. L. G. Lundsted and J. P. McMahon, U.S. Patent 2,718,509 to Wyandotte Chemicals Corporation (1955).
88. G. E. Hall, Jr., U.S. Patent 2,713,572 to Wyandotte Chemicals Corporation (1955).
89. J. S. Canaday and E. K. Moss, U.S. Patent 5,001,165 to Sloss Industries Corporation (1991).
90. Z. Funke and L. Starke, *Acta Polym. 43*:21 (1992).
91. S. A. Snow, W. N. Fenton, and M. J. Owen, *J. Cell. Plast. 26*:172 (1990).
92. L. G. Lundsted and I. R. Schmolka, in *Block and Graft Copolymerization*, Vol. 2 (R. J. Ceresa, Ed.), John Wiley & Sons, Ltd., London, 1976, pp. 114–117.
93. G. Frisch and K. Albrecht, U.S. Patent 5,074,905 to Hoeschst Aktiengesellschaft (1991).
94. C. J. Ong, U.S. Patent 5,270,286 to American Cyanamid Company (1993).
95. G. A. Policello, U.S. Patent 5,104,647 to Union Carbide Chemicals & Plastics Technology Corporation (1992).
96. J. Pinter, J. Pal, E. Kiss, E. Shuszler, S. Angyan, L. Pap, A. Szego, T. Detre, and T. Marmarosi, U.S. Patent 5,045,311 to Chinoin Gyogyszer es Vegyeszeti Termekek Gyara Rt. (1991).
97. I. R. Schmolka, U.S. Patent 4,735,015 to BASF Corporation (1988).
98. D. M. French, *J. Polymer Sci. 32*:395 (1958).
99. D. M. French, U.S. Patent 2,998,400 to Wyandotte Chemicals Corporation (1961).
100. M. Niessner, S. Wickel, W. Schneider, J. Beck, H. Hartmann, and T. Meyer, U.S. Patent 5,258,473 to BASF Aktiengesellschaft (1993).
101. M. Ara and A. Katoh, U.S. Patent 5,378,291 to Nihon Parkerizing Co., Ltd. (1995).
102. I. T. Kim and P. F. Luckham, *Colloids Surf. 68*:243 (1992).
103. B. Wasyliw, E. Stone, and J. G. Pucknat, U.S. Patent 5,284,894 to BASF Corporation (1994).
104. Y. C. Chen, V. Dimonie, and M. S. El-Aasser, *J. Appl. Polym. Sci. 46*:691 (1992).

105. Y. C. Chen, V. Dimonie, and M. S. El-Aasser, *J. Appl. Polym. Sci. 45*:487 (1992).
106. W. D. Hergeth, P. Bloss, F. Biedenweg, P. Abendroth, K. Schmutzler, and S. Wartewig, *Makromol. Chem. 191*:2949 (1990).
107. A. Damusis, J. Compton, and C. Leaf, *TAPPI 42*:413 (1959).
108. H. K. Salzberg and C. J. King, U.S. Patent 2,933,406 to The Borden Company (1960).
109. R. S. Lamar, U.S. Patent 2,844,486 to Sierra Talc & Clay Company (1958).
110. W. G. Drautz, U.S. Patent 3,404,014 to GAF Corporation (1968).
111. T. Kondo and Y. Tsubai, U.S. Patent 5,174,815 to Mitsubishi Paper Mills Limited (1992).
112. R. J. Holland, M. J. Anchor, and C. G. Utz, U.S. Patent 5,169,894. Unassigned (1992).
113. R. Bruttel and P. Kvita, U.S. Patent 4,961,755 to Ciba-Geigy Corporation (1990).
114. T. B. Gunnell, J. B. Hines, Jr., and C. N. Barry, U.S. Patent 5,331,097 to Milliken Research Corporation (1994).
115. I. Bast and M. Scholl, U.S. Patent 5,228,953 to BK Ladenburg GmbH Gesellschaft fuer chemische Erzeugnisse (1993).
116. N. R. Srivatsa, R. R. Wesolowski, and D. J. Kerstanski, U.S. Patent 5,259,969 to International Paper Company (1993).
117. Y. Nosaka and O. Ishiguro, U.S. Patent 5,248,724 to Denki Kagaku Kogyo Kabushiki Kaisha (1993).
118. R. W. Henn and D. W. Miller, U.S. Patent 3,347,675 to Eastman Kodak Company (1967).
119. K. M. Milton, U.S. Patent 3,294,537 to Eastman Kodak Company (1966).
120. G. P. Turner and D. S. Riley, U.S. Patent 4,945,030 to Horsell Graphics Industries Limited (1990).
121. A. K. Tsaur and M. Kam-Ng, U.S. Patent 5,171,659 to Eastman Kodak Company (1992).
122. A. K. Tsaur and M. Kam-Ng, U.S. Patent 5,210,013 to Eastman Kodak Company (1993).
123. P. Bagchi, G. J. McSweeney, and S. J. Sargeant, U.S. Patent 5,013,640 to Eastman Kodak Company (1991).
124. A. K. Tsaur and M. Kam-Ng, U.S. Patent 5,147,773 to Eastman Kodak Company (1992).
125. S. H. Kim, M. Kam-Ng, A. K. Tsaur, J. I. Cohen, R. A. Demauriac, G. H. Hawks III, and J. D. Baloga, U.S. Patent 5,236,817 to Eastman Kodak Company (1993).
126. K. M. Hellsten and S. O. Santamaki, U.S. Patent 3,850,831 to Mo Och Domsjo Aktiebolag (1974).

127. L. A. Gilbert, U.S. Patent 4,284,524 to The Proctor & Gamble Company (1981).
128. L. G. Lundsted and I. R. Schmolka, in *Block and Graft Copolymerization*, Vol. 2 (R. J. Ceresa, Ed.), John Wiley & Sons, Ltd., London, 1976, pp. 131–159.
129. C. J. Bunczk and P. A. Burke, U.S. Patent 5,049,299 to Kiwi Brands Incorporated (1991).
130. W. R. Camp, C. J. Bunczk, and P. A. Burke, U.S. Patent 5,043,090 to Kiwi Brands, Inc. (1991).
131. I. R. Schmolka, U.S. Patent 4,387,217 to BASF Wyandotte Corporation (1983).
132. E. J. Pancheri and M. H. Mao, U.S. Patent 5,167,872 to The Procter & Gamble Company (1992).
133. R. J. Holland, U.S. Patent 5,152,933 to BASF Corporation (1992).
134. I. I. Secemski and J. L. Lynn, U.S. Patent 5,049,303 to Lever Brothers Company (1991).
135. R. Carmello and R. D. Miller, U.S. Patent 5,008,029 to Block Drug Company, Inc. (1991).
136. J. J. Burke and R. R. Roelofs, U.S. Patent 5,126,068. Unassigned (1992).
137. T. Chang, Canadian Patent Application 2,080,396 to Block Drug Company, Inc. (1994).
138. R. Gresser, U.S. Patent 4,724,095 to Rhone-Poulenc Chimie de Base (1988).
139. A. E. Winston, K. A. Jones, F. R. Cala, A. Vinci, and M. S. Lajoie, U.S. Patent 5,261,967 to Church & Dwight Co., Inc. (1993).
140. C. Pacifico, J. G. Kramer, and R. M. Abbott, *J. Soc. Cosmetic Chemists 3*:303 (1952).
141. I. R. Schmolka, *Am. Perfumer and Cosmetics 82*:25 (1967).
142. I. R. Schmolka, *Cosmetics and Perfumery 89*:63 (1974).
143. I. R. Schmolka, *Cosmetics and Toiletries 97*:61 (1982).
144. S. Mori and C. Makino, U.S. Patent 5,035,881 to Sunstar Kabushiki Kaisha (1991).
145. S. Mori and C. Tomita, U.S. Patent 5,292,528 to Sunstar Kabushiki Kaisha (1994).
146. S. A. Nathoo, M. B. Chmielewski, and S. Fakhry-Smith, U.S. Patent 5,171,564 to Colgate-Palmolive (1992).
147. M. R. Burke, M. Prencipe, and J. M. Buchanan, U.S. Patent 5,292,502 to Colgate-Palmolive Co. (1994).
148. A. L. Bianchi, J. T. Freiberg, and K. D. Konopa, U.S. Patent 5,028,413 to Bausch & Lomb Incorporated (1991).
149. N. Nabi, A. Gaffar, and J. Afflitto, U.S. Patent 5,275,805. Unassigned (1994).

150. R. L. Mitchell and G. A. Durga, U.S. Patent 5,096,698 to Colgate-Palmolive Company (1992).
151. A. Gaffar, N. Nabi, J. Afflitto, and O. Stringer, U.S. Patent 5,294,431 to Colgate-Palmolive Co. (1994).
152. E. J. Carlin, A. K. Talwar, L. T. Principe, and S. S. Dills, U.S. Patent 5,100,650 to Warner-Lambert Company (1992).
153. K. D. Konopa, U.S. Patent 5,292,527 to Bausch & Lomb Incorporated (1994).
154. C. J. Kleber and M. S. Putt, U.S. Patent 4,976,954 to Purdue Research Foundation (1990).
155. C. W. Cardin, J. I. Davis, J. L. Hart, and D. G. Schmidt, U.S. Patent 5,104,645 to The Proctor & Gamble Company (1992).
156. I. I. Walele, N. J. Scarangella, A. Ansaldi, A. M. Andrews, and S. A. Syed, U.S. Patent 5,271,930 to Finetex, Inc. (1993).
157. J. F. Andrews and J. T. Kure, U.S. Patent 5,378,731 to Minnesota Mining and Manufacturing Company (1995).
158. S. E. Piechota, Jr., U.S. Patent 5,256,396 to Colgate-Palmolive Company (1993).
159. R. J. Lipinski, U.S. Patent 3,239,467 to Lord Corporation (1966).
160. P. D. Deck, J. B. Rivera, and W. L. Harpel, U.S. Patent 5,114,607 to Betz Laboratories, Inc. (1992).
161. J. G. Schaffhausen, U.S. Patent 5,204,012 to Ethyl Corporation (1993).
162. G. Mika, U.S. Patent 4,919,752 to Kollmorgen Corporation (1990).
163. M. Ikeda and J. Kataoka, U.S. Patent 5,269,957 to Taiho Industries Co., Ltd. (1993).
164. M. S. Dahanayake, U.S. Patent 5,045,232 to Rhone-Poulenc Specialty Chemicals, L. P. (1991).
165. Y. Sasaki and J. C. Ercillo, U.S. Patent 5,322,876 to Avery Dennison Corporation (1994).
166. M. M. Blackstone, C. Chen, and T. W. Woodward, U.S. Patent 4,906,331 to Betz PaperChem, Inc. (1990).
167. R. L. Mitchell, S. Schlosser, and P. H. Schlosser, U.S. Patent 2,999,045 to Rayonier Incorporated (1961).
168. J. N. Faulks, U.S. Patent 5,019,291 to BP Chemicals Limited (1991).
169. E. L. Morehouse, U.S. Patent 3,669,913 to Union Carbide Corporation (1972).
170. D. C. Mente, U.S. Patent 4,997,471 to BASF Corporation (1991).
171. D. C. Mente, U.S. Patent 5,124,389. Unassigned (1992).
172. *Res. Discl. 310*:168 (1990).
173. M. G. Haigh and B. V. Harrowfield, in *Proceedings of the International Wool Testing Research Conference*, 8th, Vol. 3 (G. Crawshaw, Ed.), Wool Res. Organ. N. Z., Christchurch, N.Z., 1990, pp. 277–285.

174. P. N. Hurter and T. A. Hatton, *Langmuir 8*:1291 (1992).
175. B. Ellis, P. Harold, and H. Kronberg, *Environ. Technol. 12*:447 (1991).
176. W. Klinger and E. Milewski, U.S. Patent 5,298,193 to Hoechst Aktiengesellschaft (1994).
177. R. L. Henry and R. E. Leach, U.S. Patent 5,135,751 to Mediventures Incorporated (1992).
178. M. Blackstone, U.S. Patent 5,298,120. Unassigned (1994).
179. N. Ertrin, P. Crosley, C. Haynes, (Eds.), *CTFA Cosmetics Ingredient Dictionary*, 3rd ed., The Cosmetic, Toiletry, and Fragrance Association, Inc., Washington, D.C., pp. 479, 481.
180. I. R. Schmolka, *J. Am. Oil Chemists Soc. 68*:206 (1991).
181. M. A. Faers and P. F. Luckham, *Colloids and Surfaces A: Physiochemical and Engineering Aspects, 86*:317 (1994).

6

Toxicology of Polyoxyalkylene Block Copolymers

STEPHEN C. RODRIGUEZ* Department of Toxicology, Stonybrook Laboratories Inc., Princeton, New Jersey

EDWARD J. SINGER Technical Consultant, Environmental Science & Toxicology, Belle Mead, New Jersey

I.	Introduction	212
	A. Structure	212
	B. Synthesis	213
	C. Structural variations	213
	D. Structurally-related properties	213
	E. Nomenclature	214
II.	In Vivo Mammalian Toxicology Studies	214
	A. Biological effects of short-term exposure (acute, subacute)	214
	B. Biological effects of long-term exposure	221
	C. Metabolic studies	223
III.	Other Biological Studies	223
	A. Predictive tests for mutagenicity and genotoxicity	223
	B. Other studies demonstrating, predicting, or related to effects on humans	224
IV.	Structure-Activity Relationships (SAR)	228
V.	Summary	230
VI.	Appendix	231
	References	240

* *Current affiliation*: Central Research, Rhône-Poulenc Rorer, Collegeville, Pennsylvania

I. INTRODUCTION

The structural aspects of the nonionic surfactants termed polyoxyalkylene block copolymers are covered in other portions of this monograph. However, certain features which may have pertinence to the toxicologic properties are discussed, insofar as such structure-activity relations have been defined for this class of materials.

A. Structure

1. Di- and Triblock Copolymers

The more simple polyoxyalkylene block copolymers are bicomponent polymeric entities in the A-B or A-B-A format, better shown as A_a-B_b, or A_a-B_b-A_a with the subscripts indicating the number of monomer units within each block. The *triblock* copolymers of the poloxamer series have a polyoxyethylene-polyoxypropylene-polyoxyethylene (POE-POP-POE) sequence centered on a (hydrophobic) polypropylene glycol nucleus, terminated by two primary hydroxyl groups. Commercial examples are best exemplified by the Pluronic® series of copolymers offered by BASF.

2. Reverse Triblock Copolymers

So-called "reverse" block copolymers also exist in which the triblock (POP-POE-POP) copolymers of meroxapol series are centered on a (hydrophilic) polyethylene glycol nucleus terminated by secondary hydroxyl groups. Examples include members of the Pluronic and Antarox® copolymer series.

3. Octablock Copolymers

The octablock copolymers have four A_a-B_b chains (or in reverse, four B_a-A_b chains), attached to an ethylene diamine nucleus. The copolymers of the octablock poloxamine series, $(POE_a\text{-}POP_b\text{-})^4$, are terminated by primary hydroxyl groups. This category is exemplified mainly by the Tetronic® block copolymers.

4. Reverse Octablock Copolymers

The reverse octablock copolymers, termed poloxamine-R in this chapter, have four B_a-A_b chains attached to an ethylene diamine nucleus. The copolymers of the octablock poloxamine-R series, $(POP_a\text{-}POE_b\text{-})^4$, would, like the meroxapol copolymers, be terminated by secondary hydroxyl groups. The Tetronic-R copolymers are in this category.

5. Binary-Component Chain Copolymers

A more complex type of block copolymer has three binary chains, each comprised of hydrophile (POE)-rich binary blocks linked to hydrophobe (POP)-rich binary blocks which are joined to a trifunctional alcohol such as glycerol or

$$HO(CH_2CH_2O)_a\,(\underset{\displaystyle CH_3}{\underset{|}{C}}HCH_2O)_b\,(CH_2CH_2O)_cH$$

FIG. 1 Typical polyoxypropylene-polyoxyethylene block copolymer.

trimethylol propane. The chains are terminated primarily by secondary rather than primary hydroxyl groups. These are commercially available as the Pluradot® series copolymers.

B. Synthesis

A typical triblock copolymer of the poloxamer series is synthesized by reaction of propylene oxide with a propylene glycol nucleus, followed by addition of ethylene oxide (EO) to both ends of the resulting POP glycol [1–3]. The resulting polyoxyalkylene block copolymer, or poloxamer, is of the type shown in Fig. 1. The structure and synthesis of the other types of block copolymer are discussed elsewhere in this monograph and by Schmolka [1].

C. Structural Variations

Within any particular class, of course, there is a broad spread of molecular weights, with variation in the absolute and relative amounts of the POE and POP subunits and hence in the overall hydrophilic or lipophilic character of the amphiphile. These all determine their specific physical properties, and therefore potential commercial utility, and have a variable influence on toxicologic properties in various test systems. For comparative purposes, some toxicologic information is also provided on random copolymer monols and diols which are similar to the block polymers in chemical and physical properties.

D. Structurally-Related Properties

For the POP-POE block polymers, some brief generalities can be made which are germane to the bioavailability and hence potential toxicity of the polymer [3]:

1. As the molecular weight of the hydrophobic unit (POP) increases, and/or as the percent content of the hydrophilic unit (POE) increases, viscosity increases accordingly.
2. As might be anticipated, water solubility decreases with increasing molecular weight (mw) of the hydrophobic unit (POP) at a fixed percent content of hydrophile (POE); conversely, water solubility increases with increasing percentage content of the hydrophile.

3. A third composition-related property is the palatability of the polymer—of considerable import in any animal feeding study. Palatability is increasingly greater with increasing molecular weight of the hydrophobe and increasing content of the hydrophile.

For a poloxamer, the simple interrelation between the molecular weight of the hydrophobic POP base, the percentage of hydrophilic POE in the copolymer, and the molecular weight of the copolymer, can be shown as:

$$\text{mw}_{\text{POP}} + \left[\left(\frac{\%}{100}\right)_{\text{POE}} \times \text{mw}_{\text{polymer}}\right] = \text{mw}_{\text{polymer}}$$

As with all such complex reaction products, the molecular weights are an average value having a typical Poisson distribution.

E. Nomenclature

The commercial designation of the polyoxyalkylene block copolymer can often give information as to its composition [1], but the designations vary with manufacturer. Within this chapter, to help relate the physical properties to performance in biological systems, the trade name or trivial name for the copolymer will be followed by a pair of numbers in parentheses. These numbers indicate, respectively, the hydrophile-lipophile balance (HLB) number of Griffin [4] and the overall molecular weight of the copolymer. The HLBs—which range between 0 (for the most lipophilic) and 20 (for the most hydrophilic)—have been calculated using the Griffin equation [4]:

$$\text{HLB} = \frac{\text{wt\% hydrophilic group}}{5}$$

For Pluronic L-64, for example, with a 40% content of POE and a molecular weight of 2,900, this would appear as (8/2900}. A listing of the polyoxyalkylene copolymers discussed herein, showing these properties, is included for reference as Table A1 in the Appendix.

II. IN VIVO MAMMALIAN TOXICOLOGY STUDIES

A. Biological Effects of Short-Term Exposure (Acute, Subacute)

1. Route of Administration: Oral

Acute oral toxicity of several Pluronic block polymers was evaluated following the administration of a single dose in the rat [3]. One of these copolymers was

TABLE 1 Oral LD_{50} Values (g/kg) of Pluronic Block Copolymers in Five Species[a]

Pluronic	HLB/mw	Rat	Mouse	Rabbit	Guinea pig	Dog
L-44	8/2200	5.0	—	—	—	—
L-62	4/2500	5.5	—	—	—	—
L-64	8/2900	< 5.0	—	—	—	—
F-68	16/8350	> 15.0	> 15.0	> 15.0	> 15.0	> 15.0
P-85	10/4600	> 34.6	—	—	—	—

[a] *Source*: [3,5].

also evaluated in the dog, mouse, rabbit, and guinea pig [5]. Three of the POE-POP block copolymers (Pluronic L-44, L-62, and L-64) were liquids, while two (Pluronic F-68 and P-85) were solids. The liquids were dosed neat, and the solids (Pluronic F-68 and P-85) were dosed in 50% and 25% w/v corn oil, respectively (Table 1).

Clinically, Pluronic L-44, L-62, and L-44 produced mild sedation within 2 h, which increased with dose and lengthening of observation period. Prostration and pulmonary edema with rales were noted 12 h postdosing, with severe respiratory depression prior to death. At necropsy, animals that died in the study showed marked engorgement of the lungs, distention of the stomach, and massive vascular dilatation with distention of the intestines. While Pluronic F-68 caused only mild sedation in the guinea pig and diarrhea in the dog, about 20% of the treated mice died. No effects were described for Pluronic P-85.

Four water-soluble Ucon® polymers were administered in single-dose oral intubation studies to two species of rodents and a lagomorph [6]. Two of these, Ucon 25–H-2005 (5/4100) and 75–H-1400 (10/2200), are diol-initiated random copolymers, and two, Ucon 50–HB-260 (10/960) and 50–HB-5100 (10/4000) are butanol-initiated random copolymers. The acute oral toxicity of these four copolymers are included in this discussion for comparative purposes (Table 2).

Ucon 50–HB-260, with a molecular weight of 940, was noticeably more toxic than the other three Ucon copolymers, whose molecular weights ranged from 2200 to 4100. Considering the enormity of the doses administered, and the consequent opportunities for pulmonary aspiration in the course of gastric intubation, there was no notable distinction among the remaining three materials, even though they differed in type, molecular weight, and hydrophile-lipophile balance.

2. Route of Administration: Inhalation

Several inhalation studies in animals, both acute and subchronic, have been conducted on polyoxyalkylene copolymers [7–11]. However, in only one of

TABLE 2 Acute Oral Toxicity of Ucon Water-Soluble Copolymers[a]

Test material	HLB/mw	Species	LD_{50} (ml/kg)
Random copolymer diols			
25–H-2005	5/4100	Rat, male[b]	14.1
		Rat, female[b]	35.9
		Rat, female[c]	64.0
		Mouse	22.6
		Rabbit	35.6
75–H-1400	10/2200	Rat, male[b]	64.0
		Rat, female[b]	64.0
		Rat, female[c]	16.0
		Mouse	45.2
		Rabbit	35.4
Random copolymer monols			
50–HB-260	10/960	Rat, male	6.0–7.1
		Rat, female	4.5–8.6
		Mouse	7.5
		Rabbit	1.8
50–HB-5100	10/4000	Rat, male	64.0
		Rat, female	45.2–64
		Mouse	49.4
		Rabbit	15.8

[a] *Source*: [6].
[b] Body weight between 90–120 g.
[c] Body weight between 180–400 g.

which we are aware [7], were block copolymers included. In the remaining studies cited [8–11], only random copolymers were evaluated.

In a study by Ulrich et al. [7], young Sprague-Dawley male rats were exposed, in ten 6–h sessions over a two-week period, to vapor of one of six copolymers. Three of these materials were block copolymers, three were random copolymers. The identity and some physical properties of the copolymers tested, the concentration and particle size of their vapors, and mortality data for the treated rats, are given in Table 3.

It is a remarkable observation [7] that vapors of the block and random diol copolymers elicited no toxic symptoms in rats, while under the same (or milder) conditions of exposure, vapors of the random copolymer monols caused high

TABLE 3 Mortality Data for Various Copolymers Following Inhalation Exposures[a]

Copolymer	HLB/mw	Chamber concn. (mg/m^3)	Equivalent aerodynamic diameter (μm)	Mortality	
				5 day	10 day
Random diol					
Ucon 75–H-1400	12/2350	101	2.3	0	0
Normal triblock					
Pluronic L-64	8/2900	103	1.7	0	0
Pluronic L-31	2/1100	97	1.9	0	0
Reverse triblock					
Pluronic 17R1	2/2900	102	1.8	0	0
Random monols					
Ucon 50–HB-2000	10/2900	103	2.2	60	(100)[b]
Ucon 50–HB-5100	10/4000	55	2.3	90	(100)[b]

[a] *Source*: [7].
[b] Any survivor(s) were sacrificed after first week of exposures.

mortality after as few as 3 treatments. This difference in toxicity is the more surprising because: 1) the block copolymers and random diol copolymer had HLBs spanning those of the random copolymer monols, and the molecular weights of at least two were equivalent or not greatly different; 2) the exposure concentrations of the block copolymers and random diol were equal to or greater than those of the monol copolymers, and the vapors of all materials tested were in the respirable range (< 10 μm); and 3) the basic chemical units (POE, POP) do not differ between the two types, and have identical intraunit ether linkages.

It is of interest also that the random copolymer monols showed a higher toxicity by inhalation than by other routes of administration [8]. In an inhalation toxicity study in five animal species, with a single 4–h exposure to the random copolymer Ucon 50–HB-5100 (10/4000), the rat was the most sensitive species and the dog the least [9]. As a result of nine 6–h exposures during a two-week inhalation study in the rat [10], Ucon 50–HB-5100 induced pulmonary hemorrhage at only 5 mg/m^3, and in a 13–week study [10] elicited slight pulmonary fibrosis, as well as pulmonary hemorrhage, at the lowest level tested (0.3 mg/m^3). Further, acute inhalation toxicity (inverse of the LC50) of Ucon fluids U-660, U-2000, and U-5100 increased with molecular weight at the same HLB level [11], in contrast to the general expectation that it would decrease.

To summarize this discussion, in their inhalation studies in the rat, Ulrich et al. [7] found marked differences in the response of the lung tissue, not on the basis of molecular weight as was seen in single-dose intubation studies by Smyth et al. [6], nor on the widely variant HLB properties, but on the type of copolymer, in which the random copolymer monols were rather harmful, while the random diol and block copolymers were virtually innocuous.

3. Route of Administration: Intravenous

Pluronic F-38 (Poloxamer 108) (16/5000) was used as an agent for isolation of gamma globulin for intravenous administration, carryover of which could expose a patient to 172 mg/kg in a 3–4-week treatment regimen. Pluronic F-38 was tested [12] for safety by injection via tail vein into Sprague-Dawley rats at 0.15, 1.0, 2.0, or 4.0 g/kg, five times a week for up to two weeks. Animals were sacrificed at intervals up to 14 days after the tenth injection. Heart, liver, lung, and kidney were examined by both light and transmission microscopy, and the Pluronic F-38 levels measured in serum.

No significant differences were seen in body weight gain, nor in serum chemistry (creatinine, osmolality, glutamic pyruvate transaminase) among any of the groups. Pluronic F-38 was not detectable (< 30 μg/ml) in serum 24 h after any injection. Secretion of this polymer from the tissue reservoirs into the urine was rapid, consistent with Wang and Stern's finding [13] that 94% of ^{14}C-labeled Pluronic F-38 was excreted in the urine within 3 days after intravenous injection.

In spite of the rapid excretion of the injected Pluronic F-38 and the reports that this class of material is unchanged metabolically [6,13], histologic lesions were observed in liver, lungs, and kidneys, primarily vacuolization associated with the lysosomes.*

To help interpret these observations, we might refer here to the "biological intimacy" conferred by intravenous injection, in which the usual barriers to absorption of a xenobiotic are by-passed. The histologic effects of the treatment may be due only to the physical nature of the Pluronic F-38, having nothing to do with any chemical interactions of the parent compound, or degradation product(s), with the tissue. In other words, the histologic lesions that were observed may be interpretable as an effect only of gross systemic overload by a highly hydrophilic, surface-active material.

Pluronic F-68 (16/8350), also a highly hydrophilic surfactant like Pluronic F-38, was injected intravenously in 2–month old Sprague-Dawley rats for one

* Hepatocellular vacuolization was dose dependent, with occasional necrotic hepatocytes at the 4.0 g/kg level. Vacuolization decreased during the 2–week observation time, but did not disappear. Renal tubular dilatation was seen at the high dose, but a dose-dependent vacuolization of epithelial cells in the convoluted tubules was present at 1.0 g/kg and higher. In the lung there was severe perivascular edema and interstitial congestion, but with no evidence of alveolar edema or fibrin deposition. The severity of pulmonary changes was dose-dependent and increased with time.

month in daily doses of 10 to 1,000 mg/kg [14]. Magnusson et al. found that there were no differences among groups with respect to clinical signs, food and water consumption, in hemoglobin, hematocrit, red and white cell counts, or in blood analyses of urea, total protein, bilirubin, alkaline phosphatase, or ALAT. Further, there were no significant gross postmortem findings, nor any microscopic pathology except in lungs and kidneys.

In all rats dosed at 500 or 1,000 mg/kg, foam cells were found in the alveolar spaces, or attached to the alveolar walls. The cytoplasm of these cells contained lipid droplets, phospholipids being the most prominent constituent, with neutral lipid being minor. The investigators [14] speculated that Pluronic F-68 was capable of inducing a generalized lipoidosis, and that this may have been due in part to the amphiphilic quality of this block copolymer.

Above 50 mg/kg, the kidneys displayed slight, dose-related focal cortical degenerative changes, with vacuolation of cytoplasm of the tubular epithelium. There was, however, no biochemical increase in blood urea values. Again, as with Pluronic F-38, the nature of the lesions produced suggests that they resulted from a physical overload phenomenon with a surface-active agent in the tissues affected.

In another study which included Pluronic F-68, the effects of various intravenously administered materials on liver weight and cytochromes P-450 concentrations were measured [15]. A 400 mg/kg body weight dose (10 ml/kg of 4% w/v solution) of: 1) commercial grade Pluronic F-68, or 2) silica-Amberlite-purified Pluronic F-68, was injected via tail vein into Wistar rats, with sacrifice one to seven days later.

In this study, while certain perfluorochemicals increased liver weight, or increased P-450 cytochromes concentrations, there was no effect by either parameter with either of the Pluronic F-68 preparations tested. The investigators [15] noted that the "absence of changes in liver weight in animals injected with the unpurified Pluronic solution contrasted with previous observations [16] . . . and show that this surfactant can have variable effects on this tissue." The results of Lowe and Armstrong [15] agree with the findings of Magnusson et al. [14], who found no effect of prolonged intravenous administration of Pluronic F-68 on any tissues, other than mild lesions in lungs and kidneys.

Low toxicity was seen also in other extended-duration intravenous studies. In a five-week toxicity study in Sprague-Dawley rats of an amidino mitomycin derivative using extensive evaluative procedures, there were no morphologic or clinico-pathologic changes associated with the Pluronic F-68 which was separately tested at 0.1% for use as a vehicle [17].

4. Route of Administration: Ocular

Ocular irritation by topical application of several Pluronic polyols (neat and in various aqueous dilutions) was evaluated in the New Zealand White rabbit [5].

TABLE 4 Ocular Irritation Scores of Various Triblock Copolymers in Rabbit Eyes[a]

Pluronic[b]	HLB/mw	25%	50%	75%	100%
L-44 [liquid]	8/2200	6.5 (2)[c]	12.0 (4)	20.2 (8)	20.4 (10)
L-62 [liquid]	4/2500	11.2 (6)	13.8 (6)	19.2 (9)	20.3 (10)
L-64 [liquid]	8/2900	13.8 (5)	18.2 (6)	19.8(7)	21.3 (9)
P-85 [solid]	10/4600	—	—	—	4.0 (4)

[a] *Source*: [5].
[b] [liquid, solid] = Physical form of copolymer.
[c] (n) = Day of study that irritation was no longer evident.

The Draize scores for each time period for each dilution of each material were listed; however, in order to summarize this large volume of data, only the highest Draize score and the day that irritation cleared are presented (Table 4). In all cases, the highest Draize scores were seen on Day 1 of the study.

The data from the lower dilutions (25 and 50%) suggest that: 1) irritation increases with increasing molecular weight, and 2) dilution in water decreases irritation. The 75% and neat polyols had the same amount of irritation. Pluronic P-85, with the highest molecular weight, was the least irritating of all the polyols.

Davidorf et al. [18] evaluated Pluronic F-127 (14/11500) as a vitreous humor substitute and an intraocular drug delivery system. The vitreous humor was removed from the eyes of New Zealand rabbits, and the fluid replaced with either Pluronic F-127 (20% aqueous) or a balanced salt solution. Any initial irritation subsided during the first week, and the intraocular pressure returned to normal after about 4 days. However, severe retinal injury occurred, which would preclude the use of the Pluronic F-127 solution as it was introduced in this study. Based on the nature of the histologic lesions in the retina, the investigators [18] interpreted the retinal injury as having most likely resulted from the detergent effects of the Pluronic F-127, with lysis of the lipid component of the retina.

Pluronic F-108 (16/15500) was evaluated as a potential postcorneal refractive material for use in eye surgery because of its propensity for gelation at high temperatures, while liquid at low temperatures [19]. Following injection of the Pluronic F-108 in the liquid state into a surgically prepared corneal bed in 17 rabbits, optical measurements using biomicroscopy (histology), keratometry (curvature), and pachymetry (thickness) were made at regular intervals for three months. The Pluronic F-108 was well tolerated by the cornea.

In addition to the short-term exposure toxicology data published in peer-reviewed journals, there is a large body of information developed by the manufacturers of the various polyoxyalkylene copolymers which has not been published formally. A summary of such studies, which have been made available to us, are included in Tables A2–A7 in the Appendix. No longer-term studies have been included in this tabulation, since insufficient data were available to make reasonable interpretations.

B. Biological Effects of Long-Term Exposure

1. Route of Administration: Oral

Four water-soluble Ucon fluids tested by Smyth et al. [6] in single-dose studies, were also evaluated for chronic toxicity in near life-time (two-year) feeding studies in the white rat, and in two-year feeding studies in the dog. Two of these, Ucon 75–H-1400 (10/2200) and Ucon 25–H-2005 (5/4100) are random POE-POP diol-initiated copolymers, while the other two, Ucon 50–HB-260 (10/940) and Ucon 50–HB-5100 (10/4000) are random, butoxy-initiated copolymers of oxyethylene and oxypropylene.

(a) Rat Study. In the rat study [6], the dosage ranges (g/kg/day) shown in Table 5 were used, and held constant by periodic adjustment. Female rats consuming the high dose of Ucon 25–H-2005 random diol showed depressed weight gain. Apart from this, there was no statistically valid difference between any dose group and the control groups by the parameters employed (behavior, diet consumption, mortality, life span, incidence of infections, liver-to-body weight ratio or kidney-to-body weight ratio, body weight gain, hematocrit, total red blood cell count, incidence of neoplasms, histopathology in 20 tissues).

The potential for human exposure to the four test materials is not known, but the maximum dose levels fed (0.5 g/kg/day, equivalent to about 35 g per 70 kg man per day) may be maximal based on palatability studies [3]. Because of the gustatory limitation on dose level, this study may not meet the "maximum

TABLE 5 Doses Added to the Diet for Two-Year Study in the CF-E Rats[a]

Test material	HLB/mw	No. levels	Range (mg/kg/day)	Type copolymer
25–H-2005	5/4100	3	20–500	Random diol
75–H-1400	15/2200	2	100–500	Random diol
50–HB-260	10/960	4	4–500	Random monol
50–HB-5100	10/4000	3	20–500	Random monol

[a] *Source*: [6].

tolerated dose" criterion for a carcinogenicity study. However, this is no deficiency if the weight-loss criterion would be satisfied only because the animals were semistarved as a result of an unpalatable diet.

Clearly, none of the materials have anything more than a weak carcinogenic potential, in as much as there were no significant differences in the incidence of neoplasms at the highest level dosed. Unfortunately, the publication [6] does not detail the incidence of the various tumors that were observed, and which might be expected to arise spontaneously in any such study.

(b) Dog Study. For the dog study [6], dosages are expressed in Table 6 as: 1) percent of the diet, and 2) as the average equivalent in g/kg/day (in parentheses). These dosages were held constant by periodic adjustment during the study. At the high-dose level (1.67%) of Ucon 25–H-2005, granular degeneration of the cytoplasm of the intestinal wall smooth muscle was seen, a condition termed leiomyometaplasia. The significance of this lesion is not known. No other statistically valid difference was seen between any dose group and the control group for any of the observations made: appetite, body-weight change, mortality, liver-to-body weight or kidney-to-body weight ratio, hematocrit, hemoglobin, red and white blood cell total counts, differential white cell counts, serum urea nitrogen, serum alkaline phosphatase, 15–minute BSP retention, gross pathology, or histopathology in 18 tissues.

Because 25–H-2005 random diol is the most lipophilic compound of the group, this physical property may have had some influence on both the diminished weight gain in the female rats and the appearance of the intestinal "lesion" in the dogs over the course of the treatment. Johnston and Miller [20], in a

TABLE 6 Doses Added to the Diet for Two-Year Study in the Beagle Dog[a]

Test material	HLB/mw	No. levels	Range % of diet (mg/kg/day)	Type of copolymer
25–H-2005	5/4100	3	0.067–1.67 (26–620)	Random diol
75–H-1400	15/2200	2	0.33–1.67 (130–500)	Random diol
50–HB-260	10/960	3	0.0135–1.67 (4.3–620)	Random monol
50–HB-5100	10/4000	3	0.067–1.67 (23–610)	Random monol

[a] *Source*: [7].

toxicologic evaluation of poloxamer triblock copolymers for use as vehicles for intramuscular use, concluded that the toxicity of the poloxamer vehicles to rabbit muscle was directly proportional to their lipophilicity.

2. Route of Administration: Dermal (Topical)

In a lifetime dermal carcinogenesis study in C3H/HeJ mice [6] two undiluted random diol copolymers (Ucon 75–H-1400 and 25–H-2005) were painted three times per week on the clipped back of mice (no volume specified). Methylcholanthrene (0.2%) was the positive control, acetone the negative control. While neoplasms developed in all of the methylcholanthrene-treated mice, no papillomata or carcinomata developed in the acetone- or polymer-treated mice.

C. Metabolic Studies

For determination of their metabolic fate, two random copolymer diols (Ucon 75–H-1400 and Ucon 25–H-2005) were labeled with ^{14}C in the EO groups [6]. A dose of 67 mg/kg was administered by oral intubation to rats, which were sacrificed 7 days later.

Of the 80% of ^{14}C recovered from 75–H-1400, virtually all (76%) was excreted in the feces. Of the 80% recovered from 25–H-2005, about 50% was in the feces, and 26% in the urine, indicating better absorption of the more lipophilic copolymer. The lack of any ^{14}C in the expired carbon dioxide for either of the copolymers indicated that even when there was absorption, there was no catabolism.

Wang and Stern [13] gave a single intravenous injection (7 or 100 mg/kg) to rats of ^{14}C-labeled triblock copolymer Pluronic F-38 (16/5000). The $T_{1/2}$ in plasma was found to be about 6 min; within 3 days, 94% of the ^{14}C label was excreted in the urine and 6% in the feces (probably from biliary excretion). Silica gel chromatography confirmed that only the parent compound was present in the urine, again indicating that this type of copolymer undergoes no metabolic change in the body.

III. OTHER BIOLOGICAL STUDIES

A. Predictive Tests for Mutagenicity and Genotoxicity

Evaluation of the mutagenic potential of complex hydrocarbon mixtures in the *Salmonella*/Microsomal mutagenicity assay has been an ongoing problem. Blackburn et al. [21] addressed the problem by using dimethylsulfoxide (DMSO) as an extracting agent for the active materials. Marino [22] found that Pluronic F-127 (14/11500), 50% w/v in absolute alcohol, was an effective emulsifying (solubilizing) agent for petroleum hydrocarbons. It was also found that the cytotoxicity and mutagenicity responses to Pluronic F-127 of five *Salmonella* strains (TA97, TA98, TA100, TA1535, and TA1537) were negative, and no different from those to DMSO.

Papciak et al. [23] found that Pluronic F-127 (14/11500) did not induce genotoxicity in the Chinese hamster ovary/HGPRT assay, the primary rat hepatocyte/DNA repair test, or the cell transformation assay using Balb C/3T3 cells. In another mutagenicity assay, the Drosophila Wing Spot Test, Pluronic F-68 (16/8350) was included at 0.3% in a control medium [24]. The various mutagenicity indices for the Pluronic-containing medium were no different from those seen with the water-control medium.

B. Other Studies Demonstrating, Predicting, or Related to Effects on Humans

1. Immunologic Studies

The potential of 17 different block copolymers to stimulate antibody responses (adjuvant action) and to induce tissue toxicity (inflammation) was studied by Hunter and Bennett [25]. The action of these surfactants was found to vary considerably with the relative composition (POE to POP ratio) within the normal (POE_a-POP_b-POE_c) or reverse structure (POP_a-POE_b-POP_c), as well as with the configuration of the molecules (as triblocks or octablocks). Each agent produced a distinct pattern of immune response and inflammation, as is discussed in Section IV on Structure-Activity Relations.

Pluronic L-121 (2/4500), one of the block copolymers [25] found to induce a high titer of antibody to bovine serum albumin (BSA), was incorporated in an adjuvant formulation for use in vaccines [26,27]. This formulation, which also included the threonyl analog of muramyl dipeptide, Tween 80, and squalene, was effective with several antigens, and in several species, in increasing both cell-mediated (IgE) and humoral (IgG) immunity, and was free of any significant side effects.

Histamine release from mast cells is a prime event in allergic reactions, both immediate and delayed, and is responsible for many of the manifestations of such reactions such as increased cell permeability. The (POP-POE-) reverse octablock copolymers are ionophores* for monovalent cations [28], and induce increased permeability of human erythrocytes in direct proportion to their activity as inflammatory and histamine-releasing agents in mice [29]. In further studies on reverse octablock copolymer Tetronic 130R2 (4/7200), Atkinson et al. [30] found that this class of copolymer can initiate release of a calcium-

---

* An ionophor is a compound that facilitates transmission of an ion across a lipid barrier, such as a cell membrane, either by combining with the ion or by increasing the permeability of the barrier to it.

dependent mediator (i.e., histamine) in both rodent mast cells and human basophils by mediating movement of monovalent cations.

Hence, in testing the biological effects of exposure by any route of administration, the ionophorous ability of the block copolymers, in addition to their surface active properties, should be evaluated in terms of any effects seen. Reference might be made to effects observed in the lung and kidney after intravenous injection of the triblock copolymers Pluronic F-38 and F-68 [12,14].

Interleukin 2 (IL-2) is a class of potent immunoregulatory molecules that triggers clonal expansion of sensitized or activated T-cells, which are capable of destroying tumor cells. Using Pluronic F-127 (14/11500) as a suspending agent to foster slow release of IL-2, Morikawa et al. [31] injected the preparations (either 10,000 units or 60,000 units) subcutaneously in Wistar (WKA) rats, 10 times per day, every other day from Day 1 to Day 19.

The antitumor activity of IL-2 against the transplantable rat fibrosarcoma KMT-17 was considerably enhanced by the Pluronic F-127 vehicle as compared to saline. With either of the IL-2 doses given in Pluronic F-127, 1 of the 8 rats treated survived (0 for the saline vehicle), and there was an extension of about 50% of the mean survival time for the remaining rats. The investigators suggested that the enhanced therapeutic effects of the IL-2/Pluronic F-127 preparations resulted from sustained IL-2 activity at the tumor sites.

2. Gastrointestinal Studies

Brunelle et al. [32] investigated the influence of Pluronic L-81 (2/2750) on gastrointestinal tract absorption of lipid (^{14}C-triolein). They found that a single high dose (138 mg/rat, or ≈ 1 g/kg) of this lipophilic copolymer, given by oral gavage to male Sprague-Dawley rats, caused marked inhibition of triglyceride absorption. This was due, at least in part, to a delay in hydrolysis of the administered triglyceride.

In a four-week dietary study of lipid (^{14}C-triolein) absorption, a dose level of 30 mg/rat (≈ 150 mg/kg) of Pluronic L-81, caused no maldigestion or malabsorption [32]. The test rats were able to absorb and re-esterify dietary lipid well; however, transport of these fats out of enterocytes into the intestinal lymph appeared to be delayed.

Young male swine, maintained for seven weeks on a high-fat, high-cholesterol diet supplemented with 1% by weight of Pluronic L-81, showed a reduction in plasma cholesterol levels to about half of the unsupplemented controls (226 versus 519 mg/100 ml) [32]. Triglyceride levels were about equal in both groups, but liver weight was significantly higher in the experimental swine. By light and electron microscopy, lipid droplets were present in the enterocytes, but these were free in the cytoplasm and not membrane-bound.

In a later study to evaluate the effect of Pluronic L-81 (2/2750) on lymphatic transport of lipid, Sprague-Dawley rats were fed a fat-free semisynthetic diet (with added 10% safflower oil) containing 0.5% Pluronic L-81 for three weeks or more [33]. They were then infused for 8 h via duodenal tube with an emulsion containing ^{3}H-labeled triolein, ^{14}C-labeled cholesterol, phosphatidylcholine, and sodium taurocholate, with or without Pluronic L-81; lymph was collected via a cannula in the thoracic duct.

In the Pluronic L-81–dosed rats, lymphatic outputs of triglyceride and cholesterol were greatly impaired; however, digestion and absorption of these chemicals were not adversely affected, nor was re-esterification. The depressed lipid transport in the lymphatic system appeared to be the result of interference by the Pluronic L-81 with lipoprotein assembly and/or exit of lipoproteins from the mucosal cells, perhaps interfering with synthesis of the chylomicrons necessary for such transport. This effect of Pluronic L-81 was readily reversible—the ability to transport lipid was regained within 24 h after discontinuance of Pluronic L-81 feeding.

To define the effect of surface-active agents on digestion and absorption in the small intestine, Ohsumi et al. [34] added Pluronic F-68 (16/8350) or Tween 80 to the drinking water of Wistar rats, at a level of 5% or 10%, for a period of 12 weeks. Measurements were made of glucose absorption periodically, and of sucrose and dextrin absorption terminally. Light and electron-microscopic observations were made of the intestinal tissues.

There were no remarkable changes in body-weight gain, in absorption of glucose (2% solution), sucrose (4% solution), or dextrin (2% solution), nor were there any degenerative effects on the intestinal epithelium as a result of exposure of the rats to 5% Pluronic F-68 in their drinking water. Treatment with the 10% Pluronic F-68, or 5% or 10% Tween 80 solutions resulted in depression of weight gain (10% Tween 80 $\gg$ 10% Pluronic F-68 $\approx$ 5% Tween 80), and an increase in sugar absorption, most notably in the 10% Tween 80 group. Morphological changes in these groups included deformity of mitochondria, hypersecretion of goblet cells, and shortness of the villi. While these experimental results could conceivably be of value in the formulation of pharmaceutical dosage forms, the investigators [34] might have made clearer how they related to the malabsorption syndrome.

3. Hematologic Studies

(a) Erythrocyte-Related Studies. Pluronic L-92 (4/3500) and F-127 (14/11500) were evaluated for toxicity to human erythrocytes when used as solubilizing or emulsifying agents for Amphotericin B [35]. Amphotericin B is the drug of choice for treatment of mycoses in immunosuppressed patients such as AIDS victims and transplant patients. However, in addition to its desired toxicity to fungal cells, it is also toxic to human cells.

In a 6-h assay at 38 °C, up to 60 μg/ml of Amphotericin B emulsified in egg lecithin elicited no detectable erythrocyte lysis [35]. Under the same conditions, Pluronic L-92 showed gradually increasing toxicity to a maximum of 38% lysis at 16 μg/ml, while with Pluronic F-127 cell lysis increased only to 16% over the concentration range studied. Solubilized Amphotericin B, however, elicited 100% lysis above 12 μg/ml, independent of the system evaluated.

(b) Plasma/Serum-Related Studies. Pluronic F-127 (Poloxamer 407; 14/11500), when injected intraperitoneally (i.p.) or intramuscularly (i.m.) as a 30% w/w (≈ 1.5 g/kg) solution into male Sprague Dawley rats, resulted in elevated plasma levels of cholesterol and triglyceride for greater than 96 h [36]. When commercial rat chow was ingested following i.p. injection of Pluronic F-127, there was a significantly greater elevation in plasma levels of cholesterol and triglyceride than was seen in fasted animals similarly injected.

Dose-response studies [36] indicated that synthesis of cholesterol, or its removal from systemic circulation, was more sensitive to Pluronic F-127 injection than was triglyceride. Treatment with Lovastatin (an antihypercholesterolemic drug) inhibited the elevations of either cholesterol or triglyceride induced by Pluronic F-127.

It was proposed [36] that the elevation of plasma cholesterol by Pluronic F-127 resulted from stimulation of 3–hydroxy-3–methylglutaryl-coenzyme-A reductase activity in the liver. Another paper from this laboratory [37] concluded that the predominant mechanism for the Pluronic F-127–induced increase in circulating triglyceride was a reduction in triglyceride hydrolysis due to inhibition of a lipoprotein lipase.

(c) Blood Substitute Studies. Pluronic F-68 (16/8350) was used as an emulsifier for perfluorocarbon substances tested as improved artificial blood substitutes by Yokoyama et al. [38]. Such an emulsion containing F-4–Methyloctahydroquinolidizine (FMOQ) was used for extensive transfusion of rats, which survived well for a long period, thus indicating high efficacy and low toxicity of this emulsion.

While Yokoyama et al. [38] found Pluronic F-68 satisfactory as an emulsifier in their system, artificial blood substitutes containing this same Pluronic F-68 (16/8350) have been reported to be cytotoxic to cell lines in in vitro testing. To address this situation, Lane and Krukonis [39] extracted Pluronic F-68 by supercritical fluid fractionation, using carbon dioxide under high pressure. They found that the extract was 77% more toxic and the residuum 30% less toxic to Hela cells than the parent Pluronic F-68, suggesting that such purification might be a desirable step for certain applications.

4. Oncologic Relations

It was mentioned previously that when IL-2 was suspended in Pluronic F-127 (14/11500) (as compared to saline) and administered subcutaneously, the survival time of rats with a transplantable fibrosarcoma was extended by about 50%, the enhanced therapeutic effect being attributed [31] to sustained activity of the IL-2. In vitro experiments on Pluronic F-127 gel suggested this gel might be a useful vehicle for sustained-release of mitomycin C in the peritoneal cavity; however, there was severe hepato-renal toxicity in rabbits or mice given (perhaps excessively) large amounts of the gel by the i.p. route [40].

More recently, Miyazaki et al. [41] found that Pluronic F-127 (14/11500) gels given by i.p. route enhanced the therapeutic effect of mitomycin C against a Sarcoma-180 ascites tumor in mice. The i.p. injection of mitomycin C in 25% (w/w) Pluronic F-127, one day following i.p. injection of tumor cells, prolonged the life span significantly, and more so than the free drug. The high chemotheraeutic efficiency of the mitomycin C in Pluronic F-127 at high doses was particularly striking, because at these doses the mitomycin C alone would have been toxic.

In a study focused on potentiating the antitumor action of endotoxin in mice [42], various combinations of analogs of muramyl dipeptide (MDP)—a synthetic immunoadjuvant corresponding to the smallest immunologically active glycopeptide subunit of the bacterial cell wall—and two surfactants, Pluronic L-121 (2/4500) and dimethyldioctadecylammonium bromide (DDA), were tested against solid syngeneic Meth A tumors. All agents were injected intravenously in aqueous solution. None of the muramyl peptide analogs, nor Pluronic L-121 nor DDA, had any strong antitumor action of its own. A combination of endotoxin, MDP, and Pluronic L-121 caused complete tumor regression in all mice, but was highly toxic. Again the potential role of the poloxamers as adjuvants was recognized, but could not be exploited in this system.

IV. STRUCTURE-ACTIVITY RELATIONSHIPS (SAR)

Hunter and Bennett [25] evaluated the ability of 17 surface-active agents (Pluronic L-81, L-92, L-101, L-121, L-122, 25R1, 31R1; Tetronic T-1101, T-1301, T-1501, T-90R1, T-110R1, T-130R1, T-130R2, T-150R1, T-150R4, and T-150R8), all polyoxyalkylene block copolymers, to enhance antibody formation (i.e., adjuvant activity) and elicit inflammation in the mouse footpad assay using BSA as the antigen.* These copolymers differed in molecular weight, in the "nucleus" of the copolymer, in the number and mode of linkage of the POE to the POP blocks, and in the relative number of the POE and POP units comprising each of the blocks.

Each of the 17 copolymers produced a different pattern of immune response and inflammation:

1. For copolymers with the "normal" structure (POE_a-POP_b-POE_c)

* Oil-in-water emulsions contained 1 mg BSA as antigen, 50 mg of copolymer, and 100 μl of mineral oil, in 2 ml of phosphate-buffered saline containing 0.2% Tween 80.

a. The most effective adjuvants were the largest, most hydrophobic polymers, namely, triblocks Pluronic L-121 (2/4500) and L-101 (2/3800), and octablock T1501 (2/7500);
b. As the length of the POE chains was increased, the preparations lost their activity as immune system stimulants (adjuvants) and induced less inflammation;
c. A decrease in the molecular weight of the molecule, while maintaining a fixed proportion of POP and POE, decreased adjuvant activity but increased the toxicity, as indicated by inflammation at the site of injection.

2. Copolymers with the reverse structure (POP_a-POE_b-POP_c) were weak adjuvants, but tended to stimulate granulomatous inflammation and proliferation of fibroblasts.

The relationship between HLB and adjuvant activity was not a simple one. All of the strong adjuvants had low HLB values as had been seen in an earlier study [43], but several copolymers with low HLB values had no demonstrable adjuvant activity for BSA.

Several observations indicated that a highly specific mechanism was involved. Pluronic L-121 (2/4500) stimulated a mean antibody titer to BSA of 67,000, whereas Pluronic L-122 (4/5000), similarly injected, stimulated a titer of only 184. The Pluronic L-121 and L-122 differ only in the length of the POE chains. Again, Pluronic L-101 (2/3800) and 31R1 (2/3500) produced very different immunologic and inflammatory responses (antibody titers of 84,000 versus 264), even though composed of markedly similar components, but attached in the reverse order. Among the octablock copolymers, adjuvant activity increased as the inflammatory action decreased.

Hunter and Bennett suggested [25] that much of the activity of these agents derived from their ability to form adsorptive (adhesive) surfaces on oil drops which bind and activate complement, and a variety of other biologically active molecules in addition to the antigen. If this is true, the specificity of the various block copolymers is based on their physico-chemical properties rather than on any chemical interactions, a generality which seems borne out in evaluations of the activity of these copolymers in other biological systems discussed earlier.

Johnston and Miller [20] sought to relate structurally-dependent properties of copolymers to their toxicity for muscle tissue. Such information would indicate the potential utility of these copolymers for parenteral coadministration with drugs. Four triblock poloxamers—Pluronic F-88 (Poloxamer 238; 16/ 10800), F-127 (Poloxamer 407; 14/12500), P-105 (Poloxamer 335; 10/6500), and P-123 (Poloxamer 403; 6/5750)—were injected i.m. into rabbits in single and multiple (5 daily doses) injection studies. These copolymers vary in their respective POE and POP chain lengths, and therefore in HLB and molecular weight.

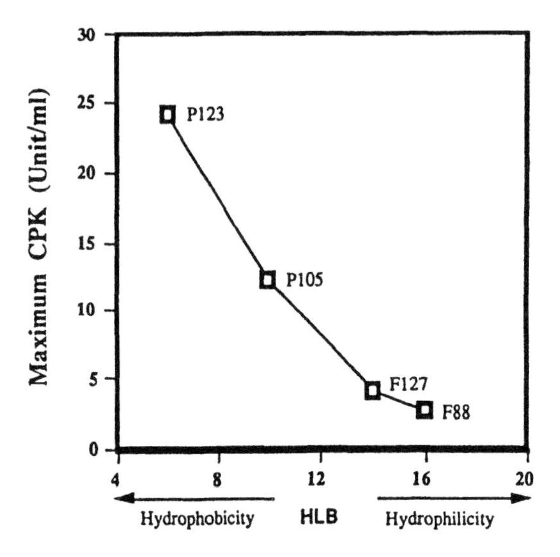

FIG. 2 Relation of creatine phosphokinase (CPK) level (a measure of muscle tissue injury) to hydrophile-lipophile balance. A HLB rating of 20 is most hydrophilic. The four copolymers tested were: Pluronic F-88 (16/10,800), Pluronic F-127 (14/12,500), Pluronic P-105 (10/6,500), and Pluronic P-123 (6/5,750). *Source*: adapted from [20].

The muscle toxicity was evaluated by gross morphological examination and also by monitoring creatine phosphokinase (CPK) level, as an index of muscle tissue injury, at various times following injection. Data adapted from their studies show the relation between the HLB and averaged maximum CPK values (Fig. 2). Clearly, the more lipophilic the copolymer (i.e., the lower the HLB), the greater the CPK value. The gross evaluations of tissue injury agreed with the serum enzyme (CPK) values after a *single* injection, somewhat less well after the *multiple* injections. Based on their data, the investigators concluded that Pluronic F-88 or F-127 (Poloxamers 238 or 407, respectively) appeared to be acceptable components for vehicles intended for i.m. use, displaying musculo-irritancy/toxicity comparable to that produced by traditional i.m. vehicles.

V. SUMMARY

These studies clearly show that as a class, block copolymers are essentially nontoxic by all routes of exposure except by intravenous injection. Any toxicity observed is the apparent result not of any chemical interaction in the normal sense of binding, but from the surface active properties determined by the various structural configurations of the POE and POP units comprising the copolymers.

Studies on the metabolic fate of the block copolymers have consistently indicated that there is no breakdown of the material, no matter what the route of administration, and that clearance via the normal excretory pathways is rather rapid. Further, whatever adverse effects have been noted were, in general, found to be readily reversible.

While the biological interactions seem to be physico-chemical, the purification study on Pluronic F-68 that disclosed a component more toxic than the parent material, should not be ignored. This could have significant implications when block copolymers are being considered for pharmaceutical formulation, particularly for dosing by injection. The extracted, more-toxic component was not defined chemically.

The Structure-Activity Relationship (SAR) studies on the structural modifiers of adjuvant action (i.e., the stimulation of antibody formation to an antigen such as bovine serum albumin) provide some of the most interesting data on the specificity of structurally-dependent interactions in biological systems. These data, as well as those relating to toxicity to muscle tissue based on the hydrophilic-lipophilic balance of the copolymer, would seem to have application in practical spheres of design and formulation of new products.

The pulmonary effects noted for the random copolymer monols may be pertinent concerning hazard evaluation in plant operations. The block copolymers and random diols have not, to our knowledge, been shown to have toxic effects by inhalation similar to those seen with some of the random copolymer monols, but, on the other hand, few have been tested for such a response.

VI. APPENDIX

The data tabulated in Tables A2–A7 are a summary of unpublished information provided by the manufacturers of various polyoxyalkylene copolymers. The studies were conducted in several contract laboratories over the span of approximately 15 years. These data were considered insufficient for analysis by structural characteristic (as triblock copolymer, reverse triblock copolymer, etc.). Therefore, while the tabulations in Tables A2–A7 are organized by structural types, a general summary is given for each route of administration. Molecular weight data were not available for the Antarox® copolymers, so they are not included in the textual summaries.

A. Route of Administration: Oral

In general, the acute toxicity of block and most random copolymers was found to be rather low in these studies, all of which were performed in the rat. Data supplied by BASF on the Pluronic and Tetronic copolymers indicate that most

TABLE A1 Commercial Name Description of Copolymers

	HLB[a]	mw[b]	Cas no.	Type of copolymer
Meroxapol 17R1	2	2,900	9003–11–6	POP-POE-POP reverse triblock
Meroxapol 25R1	2	2,800	9003-11–6	POP-POE-POP reverse triblock
Meroxapol 31R1	2	*3,500*	9003–11–6	POP-POE-POP reverse triblock
Pluronic F-38	16	5,000	9003–11–6	POE-POP-POE triblock
Pluronic F-68	16	8,350	9003–11–6	POE-POP-POE triblock
Pluronic F-88	16	10,800	9003–11–6	POE-POP-POE triblock
Pluronic F-108	16	15,500	9003–11–6	POE-POP-POE triblock
Pluronic F-127	14	*12,000*	9003–11–6	POE-POP-POE triblock
Pluronic L-31	2	1,100	9003–11–6	POE-POP-POE triblock
Pluronic L-33	6	*1,350*	9003–11–6	POE-POP-POE triblock
Pluronic L-44	8	2,200	9003–11–6	POE-POP-POE triblock
Pluronic L-62	4	2,500	9003–11–6	POE-POP-POE triblock
Pluronic L-64	8	2,900	9003–11–6	POE-POP-POE triblock
Pluronic L-72	4	2,050	9003–11–6	POE-POP-POE triblock
Pluronic L-81	2	2,750	9003–11–6	POE-POP-POE triblock
Pluronic L-92	4	3,500	9003–11–6	POE-POP-POE triblock
Pluronic L-101	2	3,800	9003–11–6	POE-POP-POE triblock
Pluronic L-121	2	*4,500*	9003–11–6	POE-POP-POE triblock
Pluronic L-122	4	5,000	9003–11–6	POE-POP-POE triblock
Pluronic L-123	6	*5,750*	9003–11–6	POE-POP-POE triblock
Pluronic P-85	10	4,600	9003–11–6	POE-POP-POE triblock
Pluronic P-105	10	*6,500*	9003–11–6	POE-POP-POE triblock
Poloxamer 108	16	5,000	9003–11–6	See Pluronic F-38
Poloxamer 238	14	11,500	9003–11–6	See Pluronic F-88
Poloxamer 335	10	*6,500*	9003–11–6	See Pluronic P-105
Poloxamer 403	6	*5,750*	9003–11–6	See Pluronic P-123
Poloxamer 407	14	*12,000*	9003–11–6	See Pluronic F-127

TABLE A1 (Continued)

	HLB[a]	mw[b]	Cas no.	Type of copolymer
Tetronic 1101	2	5,300	9003–11–6	POE-POP-POE octablock
Tetronic 1301	2	6,400	9003–11–6	POE-POP-POE octablock
Tetronic 1501	2	7,500	9003–11–6	POE-POP-POE octablock
Tetronic 90R1	2	*4,300*	9003–11–6	POP-POE-POP reverse octablock
Tetronic 110R1	2	*5,300*	9003–11–6	POP-POE-POP reverse octablock
Tetronic 130R1	2	*6,400*	9003–11–6	POP-POE-POP reverse octablock
Tetronic 130R2	4	*7,200*	9003–11–6	POP-POE-POP reverse octablock
Tetronic 150R1	2	*7,500*	9003–11–6	POP-POE-POP reverse octablock
Tetronic 150R4	8	*11,250*	9003–11–6	POP-POE-POP reverse octablock
Tetronic 150R8	16	*33,750*	9003–11–6	POP-POE-POP reverse octablock
Ucon 25–H-2005	5	4,100	9003–11–6	POE-POP random diol
Ucon 50–HB-2000	10	2,900	9038-95–3	BuO-POE-POP random monol
Ucon 50–HB-260	10	960[c]	9038-95–3	BuO-POE-POP random monol
Ucon 50–HB-5100	10	4,000	9038-95–3	BuO-POE-POP random monol
Ucon 50–HB-660	10	1,700	9038-95–3	BuO-POE-POP random monol
Ucon 75–H-1400	15	2,350[c]	9003–11–6	POE-POP random diol

[a] HLB is hydrophile-lipophile balance on a scale from 20 (most hydrophilic) to 0 (most lipophilic). The HLB was calculated using the Griffin equation (i.e., the percentage of POE in the polymer divided by 5). See [4].

[b] Italic copolymer molecular weights (mw) were calculated from the mw of the POP and the percentage of POE in the molecule.

[c] The number is an average of two differing values.

TABLE A2 Acute Toxicity of Pluronic Triblock Copolymers (Poloxamers)

		Systemic toxicity (g/kg)			Topical irritation	
	HLB/mw	Oral	Dermal	IV-Mouse	Ocular	Skin
F-38	16/5000	—	—	> 5	—	—
F-68 CS	16/8350	> 15	> 20	> 5	0.7	—
F-77		—	—	4.2	—	—
F-87		> 5	—	3.75	—	—
F-98		—	—	> 5	—	—
F-127	14/11,500	> 15	> 2	2.25	13	0.6
		> 10	> 5	—	0.0[a]	—
L-31	2/1100	—	—	0.075	—	—
		—	—	0.83	1.7	0.7
L-42		—	—	0.2	—	—
L-43		—	—	0.33	—	—
L-44	8/2200	—	—	0.75	20.4[b]	—
		—	—	—	20.2[c]	—
		—	—	—	12[d]	—
		—	—	—	6.5[e]	—
L-61		1.72	—	0.12	—	—
L-62	4/2500	5	—	0.13	20.3[b]	—
		—	—	—	19.2[c]	—
		—	—	—	13.8[d]	—
		—	—	—	11.2[e]	—

TABLE A2 (Continued)

		Systemic toxicity (g/kg)			Topical irritation	
	HLB/mw	Oral	Dermal	IV-Mouse	Ocular	Skin
L-64	8/2900	—	—	0.35	21.3[b]	—
		—	—	—	19.8[c]	—
		—	—	—	18.2[d]	—
		—	—	—	13.8[e]	—
L-72	4/2050	—	—	0.16	—	—
L-81	2/2500	—	—	0.2	—	—
L-92	4/3500	—	—	0.6	2	1.6
		—	—	0.14	—	—
L-122	4/5000	—	—	0.8	1.7	1.6
P-65		—	—	0.83	—	—
P-75		—	—	0.35	—	—
P-84		—	—	0.4	—	—
P-85	10/4600	> 35	—	0.53	1.6	—
P-94		—	—	0.6	—	—
P-103		—	—	1.4	—	—
P-104		—	—	0.75	—	—
P-105	10/6500	—	—	3	—	—
P-123		—	—	2.7	—	—

[a] 20% aqueous gel tested.
[b] Neat material tested.
[c] 75 % aqueous solution tested.
[d] 50 % aqueous solution tested.
[e] 25 % aqueous solution tested.

TABLE A3 Acute Toxicity of Pluronic Reverse Triblock Copolymers (Meroxapols)

		Systemic toxicity (g/kg)		Topical irritation	
	HLB/mw[a]	Oral	Dermal	Ocular	Skin
10R5	*10/2000*	> 28	> 10	15	0.6
10R8	*16/5000*	> 15	> 10	7.7	0.8
17R1	2/2900	2.7	> 10	7.0	0.5
17R2	*4/2125*	4.15	—	—	—
17R4	*8/2800*	> 15	> 0.9	11	0.8
17R8	*16/8500*	> 17.6	> 10	15.2	0.0
25R1	2/2800	11	> 10	6	0.9
25R2	*4/3125*	10.2	> 7	10	0.8
25R4	*8/4200*	> 15	> 10	8.2	1.0
25R8	*16/12500*	> 23	> 10	12.2	0.3
31R1	2/3500	> 34	> 10	12	0.4
31R4	*8/5200*	> 23	10.3	12	0.7

[a] Italic copolymer molecular weights (mw) were calculated from the mw of the POP and the percentage of POE in the molecule.

oral LD_{50} values are greater than 10 g/kg, and several are greater than 20 g/kg. Clinically, these materials produced diarrhea, hypoactivity, muscle weakness, and urinary incontinence, while intestinal hemorrhage and gastroenteritis were the most common findings at necropsy. Physical overloading of the animals may well have been responsible for the observed effects.

Four of these surfactants (Pluronic L-61, 17R1, 17R2, and Tetronic 701) with molecular weights below 3,100 had LD_{50} values below 5 g/kg. One was a triblock copolymer, two were reverse triblocks, and the fourth was an octablock. Tremors, ataxia, lung hemorrhage, pale liver, pale spleen, and gastrointestinal irritation were observed. However, because other structurally similar copolymers were considerably less toxic, one may wonder if pulmonary aspiration during dosing was the cause of the observed toxicity.

B. Route of Administration: Dermal (Topical)

Dermal application of block copolymers resulted in LD_{50} values ranging from < 2 g/kg to > 20 g/kg body weight in the New Zealand White rabbit. Dose-related decreases in body weight were reported following dosing with two of the

TABLE A4 Acute Oral Toxicity of Ucon Random Copolymers

	HLB/ mw	Rat (M) 90–120 g	Rat (M) 120–170 g	Rat (F) 90–120 g	Rat (F) 300–400 g	Mouse (F) 23–36 g	Rabbit (M) 2.2–2.9 kg
Diol-initiated random copolymers							
25–H-2005	5/4100	14.1[a]	—	35.9	64	22.6	35.6
75–H-1400	15/2200	64	—	64	16	45.2	35.4
Random copolymer monols							
50–HB-260	10/940	7.07	5.95	8.57	4.49	7.46	1.77
50–HB-5100	10/4000	64	—	64	45.2	49.4	15.8

[a] All acute oral toxicities are expressed in g/kg body weight.

TABLE A5 Acute Toxicity of Tetronic Copolymers (Poloxamines, R-Poloxamines)

		Systemic toxicity		Topical irritation	
	HLB/mw[a]	Oral (g/kg)	Dermal (g/kg)	Ocular	Skin
Octablock copolymers					
701	*2/3050*	3.4	> 10.2	20	3.1
908	*16/18,750*	> 23	> 10.2	17	1.8
1101	2/5300	18.8	> 10.2	11	1.6
1107	*14/15,800*	> 15	> 3	5.3	0.4
1302	*4/7200*	> 34	> 10.2	7.7	0.4
1304	*8/9600*	> 15	> 10.2	0	1.4
1307	*14/19,200*	> 15.3	> 10.2	6	0.1
1504	*8/11,250*	> 15	> 3	12	0.3
1508	*16/33,750*	> 15.3	> 3	13	0.6
Reverse octablock copolymers					
50R8	16/	> 10	> 4.6	1	0.4
70R1	2/	> 10	> 4.6	2	0.3
90R4	8/6500	> 10	> 4.6	2	0.1
150R1	2/7500	> 10	> 4.6	5.3	0.21
150R8	16/33,750	> 10	> 4.6	1.2	0.17

[a] All acute oral toxicities are expressed in g/kg body weight.

TABLE A6 Acute Toxicity of Antarox Block Copolymers

	HLB/mw	Systemic toxicity oral (g/kg)	Topical irritation	
			Ocular	Skin
LF-222	N/A	1.99	0	2.0
BL-214	N/A	4.1	23	2.28
BL-225	N/A	3.8 (ml?)	33.3	1.75
BL-240	N/A	2.4 ml	—	0.83

TABLE A7 Guinea Pig Sensitization

	HLB/mw	Method	
		Unknown	Buehler
Pluronic			
25R2	4/3000		Negative
31R1	2/3500		Negative
F-87			Negative
L-31	2/1100		Positive[a]
L-64	8/2900	Negative	
Tetronic			
1107	14/8075	Negative	
1508	16/12,150		Negative
50R8	16/9000		Negative
90R4	8/12,600		Negative
150R1	2/16500		Negative
150R8	16/27,000		Negative

[a] Mild on primary challenge.

triblock copolymers, although the LD_{50} was reported to be > 10 g/kg. A third triblock copolymer had a reported LD_{50} value of > 0.9 g/kg with delayed deaths from 6 to 11 days after dosing. Unfortunately, information on mortality and clinical signs of toxicity were absent from the reports, making additional conclusions impossible.

The only observation made following application of octablock copolymers was irritation of the test site. In general the block copolymers, while somewhat irritating, were not systemically toxic by dermal application, hence no correlation of toxicity with molecular weight could be made.

C. Route of Administration: Intravenous

Intravenous LD_{50} determinations were performed in mice only on the triblock copolymers. Copolymers with molecular weights below 5,000 tended to have LD_{50} values in a range below 830 mg/kg, while copolymers with molecular weights of 5,000 and above had LD_{50} values above 2,250 mg/kg. This statement has to be qualified: molecular weights were not available for all copolymers. Clinically, dyspnea was observed following treatment with these copolymers, accompanied by convulsions after dosing with several of the more toxic ones.

D. Route of Administration: Ocular

The ocular irritation data listed in Tables A2–A7 are maximum Draize scores. Generally, the block copolymers were only slight irritants when instilled into the eye of the rabbit. The mean Draize scores (with a range of 0–110) were for the most part below 20. Such scores suggest that the treatment resulted only in conjunctival irritation (redness, swelling, and discharge) without the more serious involvement of the cornea or iris. However, three of the triblock copolymers (Pluronic L-44, L-62, L-64) led to Draize scores slightly above 20, indicating some corneal involvement. The scores decreased rapidly within two days, implying that the corneal irritation was slight, with no persistent injury.

E. Route of Administration: Dermal

The skin irritation studies tabulated in Tables A2–A7 were all conducted in rabbits before 1983 and, most probably, follow Federal Hazardous Substance Act (FHSA) guidelines which required a 24-hour semi-occluded exposure on both intact and abraded skin. The skin irritation data provided are Primary Irritation Index (PII) scores calculated from irritation readings at 24 and 72 h post-dosing.

Only slight skin irritation was reported for any of the copolymers except for one of the octablock copolymers (Tetronic 701) which was moderately irritating. None of the copolymers tested would have required labeling as an irritant under FHSA guidelines.

REFERENCES

1. I. R. Schmolka, *J. Am. Oil Chem. Soc. 54*:110 (1977).
2. I. R. Schmolka and R. E. Seizinger, *J. Am. Oil Chem. Soc. 45*:715 (1968).
3. C. W. Leaf, *Soap & Chem. Spec. Aug.*:48 (1967).
4. W. C. Griffin, *Soc. Cos. Chemists 5*:249 (1954).
5. *Technical Brochure:Pluronic® Polyols Toxicity and Irritation Data*, 3rd Ed., Rev., BASF Wyandotte Corp, MI, 1971.
6. H. F. Smyth, C. S. Weil, J. M. King, J. B. Knaak, L. J. Sullivan, and C. P. Carpenter, *Toxicol. Appl. Pharmacol. 16*:675 (1970).
7. C. E. Ulrich, R. G. Geil, T. R. Tyler, and G. L. Kennedy, Jr., *Drug Chem. Toxicol. 15*:15 (1992).
8. D. R. Klonne, D. J. Nachreiner, D. E. Dodd, P. E. Losco, and T. R. Tyler, *Fundam. Appl. Toxicol. 9*:773 (1987).
9. G. M. Hoffman, P. E. Newton, W. C. Thomas, H. A. Birnbaum, and G. L. Kennedy, Jr., *Drug Chemical Toxicol. 14*:243 (1991).
10. D. R. Klonne, D. E. Dodd, P. E. Losco, C. M. Troup, and T. R. Tyler, *Fundam. Appl. Toxicol. 10*:682 (1988).
11. D. R. Klonne, H. D. Burleigh-Flayer, D. E. Dodd, and P. E. Losco, *Inhal. Toxicol. 5*:189 (1993).
12. C. D. Port, P. J. Garvin, and C. E. Ganote, *Toxicol. Appl. Pharmacol. 44*:401 (1978).
13. A. J. Wang and I. J. Stern, *Drug Metabol. Disposit 3*:536 (1975).
14. G. Magnusson, T. Olsson, and J-A. Nyberg, *Toxicol. Lett. 30*:203 (1986).
15. K. C. Lowe and F. H. Armstrong, *Adv. Exp. Med. Biol. 277*:267 (1990).
16. P. K. Bentley, S. S. Davis, O.L. Johnson, K. C. Lowe, and C. Washington, *J. Pharm. Pharmac. 41(9)*:661–663 (1989).
17. C. L. Bregman, C. R. Comereski, R. A. Buroker, R. S. Hirth, H. Madissoo, and G. H. Hottendorf, *Fundam. Appl. Toxicol. 9*:90 (1987).
18. F. H. Davidorf, R. B. Chambers, O. W. Kwon, W. Doyle, P. Gresak, and S. G. Frank, *Retina 10*:297 (1990).
19. J. P. Kim, R. L. Peiffer, and R. E. Holman, *J. Cataract Refract. Surg. 14*:312 (1988).
20. T. P. Johnston and S. C. Miller, *J. Parenter. Sci. Technol. 39*:83 (1985).
21. G. R. Blackburn, R. A. Deitch, C. A. Schreiner, M. A. Mehlman, and C. R. Mackerer, *Cell. Biol. Toxicol. 1*:40 (1984).
22. D. J. Marino, *Environ. Mutagen. 9*:307 (1987).
23. R. J. Papciak, S. Brecher, M. Grant, S. H. Khan, and D. J. Marino, *Environ. Mutagen. 7(Suppl. 3)*:80 (1985).
24. U. Graf, H. Frei, A. Kägi, A. J. Katz, and F. E. Würgler, *Mutat. Res. 222*:359 (1989).
25. R. L. Hunter and B. Bennett, *J. Immunol. 133*:3167 (1964).

26. A. C. Allison and N. E. Byars, *J. Immunol. Methods 95*:157 (1986).
27. N. E. Byars and A. C. Allison, *Vaccine 5*:223 (1987).
28. T. P. Atkinson, J. O. Bullock, T. F. Smith, R. E. Mullins, and R. L. Hunter, *Am. J. Physiol. 254*:C20 (1988).
29. T. P. Atkinson, T. F. Smith, and R. L. Hunter, *J. Immunol. 141*:1302 (1988).
30. T. P. Atkinson, T. F. Smith, and R. L. Hunter, *J. Immunol. 141*:1307 (1988).
31. K. Morikawa, F. Okada, M. Hosokawa, and M. Kobayashi, *Cancer Res. 47*:37 (1987).
32. C. W. Brunelle, W. J. Bochenek, R. Abraham, D. N. Kim, and J. B. Rodgers, *Am. J. Dig. Dis. 24*:718 (1979).
33. P. Tso, J. A. Balint, and J. B. Rodgers, *Am. J. Physiol. 239*:G348 (1980).
34. T. Ohsumi, M. Yonezawa, T. Aizawa, S. Motegi, F. Kawamura, M. Ohki, K. Takanashi, H. Mitsui, S. Nakagawa, and M. Kobayashi, *Nihon Univ. J. Med. 22*:39 (1980)
35. D. Forster, C. Washington, and S. S. Davis, *J. Pharm. Pharmacol. 40*:325(1988).
36. Z. G. M. Wout, E. A. Pec, J. A. Maggiore, R. H. Williams, P. Palicharla, and T. P. Johnston, *J. Parenter. Sci. Technol. 46*:192 (1992).
37. T. P. Johnston and W. K. Palmer, *Biochem. Pharmacol. 46*:1037 (1993).
38. K. Yokoyama, R. Naito, Y. Tsuda, C. Fukaya, M. Watanabe, S. Hanada, and T. Suyama, *Prog. Clin. Biol. Res. 122*:189 (1983).
39. T. A. Lane and V. Krukonis, *Transfusion 28*:375 (1988).
40. T. Abe, M. Sasaki, H. Nakajima, M. Ogita, H. Naitou, A. Nagase, K. Taguchi, and S. Miyazaki, *Gan to Kagaku Ryoho 17*:1546 (1990).
41. Miyazaki, Y. Ohkawa, M. Takada, and D. Attwood, *Chem. Pharm. Bull. Tokyo 40*:2224 (1992).
42. N. Bloksma, F. M. Hofhuis, and J. M. Willers, *Cancer Immunol. Immunother. 19*:205 (1985).
43. R. L. Hunter, F. Strickland, and F. Keady, *J. Immunol. 127*:1244 (1961).

7

Biological Activity of Polyoxyalkylene Block Copolymers in the Environment

ROBERT E. BAILEY* Environmental Toxicology Laboratory, The Dow Chemical Company, Midland, Michigan

I.	Introduction	244
II.	Environmental distribution and movement	245
	A. Water solubility	245
	B. Adsorption	245
	C. Volatilization	246
	D. Environmental distribution	246
III.	Aquatic toxicity	247
IV.	Biodegradation	249
	A. Analytical methods	249
	B. Test systems	250
	C. Substances tested	251
	D. Mechanisms	254
	References	255

* *Current affiliation*: Bailey Associates, Midland, Michigan

I. INTRODUCTION

There is little published information on the environmental fate and effects of the polyoxyalkylene block copolymers formed from ethylene oxide (EO), propylene oxide (PO) and 1,2–butylene oxide (BO). They are used not only as surfactants in the formulation of cleaning products but also in many other applications, such as foam control agents and dispersants. Thus, they have been studied for their environmental fate and effects to some extent. There are many similarities between polyoxyalkylene block copolymers and polyoxyalkylene homopolymers. It is, therefore, possible to extrapolate from the more extensive experimental data for polyoxyethylene (POE) and polyoxypropylene (POP) to predict many aspects of the environmental behavior of the block copolymers.

Aerobic biodegradation of POE has been studied in a variety of situations. Polyoxyethylene products with a molecular weight (mw) above 400 are generally not readily biodegradable in simple screening tests with low concentrations of microorganisms [1–3]. However, in acclimated systems and with higher concentrations of microorganisms, POE products with molecular weights of up to many thousands are biodegradable over a few days [4–9]. "Acclimated systems" refers to microbial populations that have been exposed to the test substance to provide an opportunity for the appropriate enzyme and population shifts to occur. Polyoxypropylene products are generally more slowly biodegraded than POE products [10]. Kawai et al. [11] isolated bacteria from soil that can live on POP (4000 mw) as a sole carbon source.

Part of the reason for the relatively small amount of research on the polyoxyalkylene products is their generally low toxicity to aquatic and terrestrial organisms. For example, POE and POP products typically have LC_{50} values for aquatic organisms of thousands of mg/L [12]. The polyoxyalkylene block copolymers are reported to have low toxicity [13] and little other toxicity information is available.

The environmental distribution and movement of the polyoxyalkylene oxides will be discussed followed by their behavior in each compartment of the environment. In this chapter, the different products will be described by their molecular weight and their percent POE content. Where more detailed information on the structure, block sizes, or other aspects of the structures are available, they will be included to facilitate interpretation of the environmental information.

II. ENVIRONMENTAL DISTRIBUTION AND MOVEMENT

A. Water Solubility

The polyoxyalkylene block copolymers are generally soluble in water. Their detailed behavior in water is generally a function of their POE content. Like other surfactants, these products are generally dispersable in water over a wide range of conditions. While the properties of the polyoxyalkylene block copolymers in water solution will not be covered in detail here, more information can be found in Chapter 3 on physical chemistry.

B. Adsorption

No published information on the adsorption of polyoxyalkylene block copolymers to soil or sediments in the environment was found. Little adsorption to soil would be expected on the basis of the generalized lipophilic interactions between soil organic matter and the polymers because of their water solubility [14], as noted above. However, polyoxyethylene has been reported to adsorb strongly to sediments with a Langmuir adsorption constant, K_d, of up to 336 mL/g for POE 1000 on a high clay content sediment [15]. The soil adsorption of POE correlated with clay content rather than organic carbon content, further indicating electrostatic or polar interactions, and the interpretation is complicated by pH and other effects. The relatively strong binding of POE is due to multiple possible binding sites on each molecule [15,16], even though each binding site may be weak. The multitude of binding sites on each molecule makes it very difficult to remove the substance by dilution because all sites must simultaneously disengage to go into solution. Thus, higher molecular weight products adsorb more strongly, which is consistent with observations.

The soil adsorption of polyoxyalkylene block copolymers is expected to be similar to or greater than that of POE with perhaps more involvement with the soil organic matter due to the lipophilicity of the POP segments. Polyoxypropylene of greater than 1000 mw is reported to be only slightly soluble in water, while POE of all molecular weights is soluble in water. The estimated log octanol/water partition coefficient, log K_{OW}, of the POE portion of a polyoxyalkylene block copolymer are close to 1 with a contribution of only 0.065 from each C_2H_4O unit [17]. Each C_3H_6O unit contributes 0.37 to the log K_{OW}. Thus it would be possible to calculate a high log K_{OW} for a polyoxyalkylene block copolymer suggesting the possibility of strong lipophilic soil adsorption in addition to the electrostatic binding.

A related phenomenon for many compounds in the environment is bioconcentration, the increased concentration of a substance in the fatty tissues of aquatic organisms. The bioconcentration factor correlates well with the K_{OW} of many compounds. However, the high molecular weight of the polyoxyalkylene block copolymers, greater than 600, as well as the hydrophilic portion is expected to prevent appreciable bioconcentration [18].

C. Volatilization

The vapor pressures of the polyoxyalkylene block copolymers are very low, much less than 1 Pa (0.008 torr). The high water solubilities and very low vapor pressures of the polyoxyalkylene block copolymers lead to a negligible level of volatilization of these compounds in the environment [19]. The very small amounts that might be present in the atmosphere as vapor would react relatively rapidly with hydroxyl radicals as a volatile organic chemical. The rate of reaction with hydroxyl radicals can be estimated using the procedure of Atkinson [20] for any combination of components of a polyoxyalkylene polymer. The reaction rate, K_{OH}, is additive for each unit and can be approximated as 13.2×10^{-12} cm^3 molecule^{-1} sec^{-1} for an EO unit and 21.2×10^{-12} cm^3 molecule^{-1} sec^{-1} for each PO unit. The reaction rate for any given polymer could be obtained by summing the contributions for each unit. Thus, the atmospheric half-life for a polyoxyalkylene compound in the atmosphere could be estimated as a fraction of a day, at an atmospheric concentration of hydroxyl radicals of 5×10^5 ·OH radicals/cm^3. This concentration of hydroxyl radicals is a diurnally, globally, and seasonally averaged value [21]. As an example, a compound consisting of 40% EO and 60% PO with a total molecular weight of 2000 would have approximately 18 EO units and 21 PO units. The hydroxyl radical reaction rate can be estimated as:

$$18(13 \times 10^{-12}) + 21(21 \times 10^{-12}) = 675 \times 10^{-12}\ \text{cm}^3/\text{molecule sec}$$

$$t_{1/2} = \frac{0.69}{K_{OH}\,[\cdot\text{OH}]} = \frac{0.69}{(675 \times 10^{-12}\ \text{cm}^3/\text{molecule sec})\ (5 \times 10^5\ \text{molecule/cm}^3)}$$

$$t_{1/2} = 2{,}053\ \text{sec} = 34\ \text{min}$$

The impact of polyoxyalkylene polymers on air quality is expected to be negligible relative to naturally occurring and automotive emissions because of the extremely small quantities that could be realistically emitted.

D. Environmental Distribution

An efficient way to approach the environmental behavior of a substance is to calculate its hypothetical distribution in the environment after release based on its physical properties, as illustrated by Mackay et al. [19]. The rates of movement and degradation in the different environmental compartments can be incorporated into the model to develop a generic representation of the movement and fate of a chemical in the environment. This type of approach is being used in some of the European risk assessment exercises which are in progress in 1995, such as the Dutch Uniform System for the Examination of Substances [22].

The physical properties assumed for this illustration of the potential distribution and movement of a polyoxyalkylene block copolymer are:

Vapor pressure	1×10^{-3} Pa
Water solubility	10,000 mg/L
Log octanol/water partition coefficient	$\log K_{OW} = 3.5$*

Putting these values into the Mackay Level One model [19] to examine the equilibrium distribution of these products indicates the majority of the product (67%) is expected to be in the water phase, with 33% adsorbed to the soil and sediments and a negligible amount in the atmosphere. Thus, the important environmental fate and effects are expected to be in the aqueous phase and associated sediments and soils.

III. AQUATIC TOXICITY

A wide range of toxicities are reported for the polyoxyalkylene products. Polyoxyethylene and POP are "practically nontoxic" to aquatic organisms in a classification scheme used by the U.S. Environmental Protection Agency [23]. That is, the LC_{50} concentrations are greater than 100 mg/L for fish and daphnids. Table 1 lists the LC_{50} values observed for a range of products. The polyoxyalkylene block copolymers have a range of reported toxicities as shown in Tables 1 and 2.

The most extensive series of toxicity tests on the polyoxyalkylene block copolymers are those published are those by Karpinska-Smulikowska [24] (Table 2). She also studied the toxicity of the same series of polymers to bacteria, *Sphaerotilis natans*, and an alga, *Scenedesmus quadricauda*. The concentrations used ranged from 0.1 mg/L to 10,000 mg/L. The results are difficult to interpret, as some of the compounds are reported to show effects for the lowest concentrations with similar effects up to high concentrations. For example, the toxicity of a 20% EO, 80% PO product (2120 mw) to *S. natans* was reported to reduce the biomass to 65% of the control at 0.1 mg/L, 47% at 1

---

* $\log K_{OW} = 3.5$ was taken to yield an estimated sediment adsorption coefficient equal to that observed by Podoll et al. [15] for POE.

TABLE 1 Toxicity of Polyoxyalkylene Homo and Block Copolymers to Aquatic Organisms

Compound	Molecular weight	Species[a]	Exposure time	LC_{50} mg/L	Ref.
POE	8000	FHM	96 h	73,493	12
POE	8000	DM	48 h	35,252	12
POE	400	FHM	96 h	87,209	12
POE	400	DM	48 h	53,484	12
POP	400	FHM	96 h	> 100	12
POP	4000	FHM	96 h	> 100	12
Triol glycerin with 12% EO + PO	3600	GO	48 h	5600	12
Triol glycerin with EO and PO	4800	FHM	96 h	4220	12
EO-PO Block copolymer		Fish	??[b]	> 100	13
EO-PO Block copolymer		Daphnid	??[b]	> 100	13

[a] FHM = fathead minnow (*Pimephales promelus*), DM = *Daphnia magna* Straus, GO = golden orf (*Leuciscus idus orfus*).
[b] Time not stated in article.

TABLE 2 LC_{50} Values (mg/L) for Polyoxyalkylene Products to Aquatic Organisms

	Species tested		
Test substance	*Paramaecium caudatum*, 48 hr	*Daphnia magna*, 48 hr	*Lebistes reticulatus*, 96 hr
41% EO, 59% PO (650 mw)	1.22	0.39[a]	41.8
40% EO, 60% PO (2060 mw)	0.35	3.56[a]	26.6
20% EO, 80% PO (2120 mw)	0.37	0.49	7.0
38% EO, 62% PO (2750 mw)	1.25	5.12	18.2
79% EO, 21% PO (5140 mw)	3.4	10.24	5.71

[a] 72 hour values.
Source: [24].

mg/L, 35% at 10 mg/L, 35% at 100 mg/L, 84% at 1000 mg/L and 302% at 10,000 mg/L. Enhanced growth as well as inhibition of both the bacteria and the alga are reported for the series of products tested. The original publication should be consulted for more details. Her results on the daphnids and fish (guppy) appear to indicate a higher level of toxicity than other reports. For example, Schoeberl [13] reports an LC_{50} value greater than 100 mg/L for an undefined product with a daphnid and Karpinska-Smulikowska [24] reports an LC_{50} value of 10.24 mg/L for the least toxic product.

IV. BIODEGRADATION

The biodegradation of polyoxyalkylene block copolymers has been studied because of the concern about discharge of surfactants used in cleaning to wastewater treatment systems and subsequent flow into surface waters. These products can also enter the environment in small amounts from many of their other applications such as synthetic lubricants. The ease of biodegradation of a compound can be tested in a variety of ways. For surfactants in general, which have been the objects of a great deal of study over the past 35 years, there are screening tests which are easy to use that employ nonspecific analyses and simple equipment such as a shaken flask. Alternatively, more elaborate studies which are designed to simulate the operation of a wastewater treatment plant can be used where the effects of foaming and adsorption on the biomass can be observed with semicontinuous or continuous operation for a period of up to 2 months. Swisher reviewed the biodegradation of all types of surfactants, including these block copolymers [25]. Biodegradation testing of materials such as polyoxyalkylene block copolymers is composed of three parts: the analytical method selected, the test system, and the substance being tested. The analytical method is chosen depending on the question to be answered or the resources available. The question may be "Does the compound act as a surfactant after treatment?". Alternatively, the question may be "Is this product completely mineralized to carbon dioxide and water?". Other methods of analysis determine the loss of soluble organics or formation of intermediate metabolites.

A. Analytical Methods

The polyoxyalkylene block copolymers are often monitored by their formation of an insoluble complex with a modified Dragendorff reagent, a mixture containing bismuth and potassium iodides and barium chloride in acetic acid [26–28]. Since a degraded sample of polyoxyalkylene block copolymer in a sewage mixture commonly contains a variety of materials, biodegradation is often reported simply as reduction of bismuth active substances (BIAS). The amount of orange-red precipitate can be determined by volume after centrifugation [26] or by potentiometric titration of the bismuth in the precipitate after filtration and dissolution in an ammonium tartrate solution [27,28]. The same reaction mixture has also been used for visualizing the polyoxyalkylene compounds in thin layer or paper chromatographic analysis of biodegradation mixtures [29]. This method is reported to be satisfactory for alkyl and nonylphenol ethoxylates containing 4 or more oxyethylene units [27].

Reduction in dissolved organic carbon (DOC) is commonly used as a measure of biodegradation in the Organization for Economic Cooperation and Development (OECD) test methods [30]. The test substance is incorporated into the microorganisms and converted into intermediate metabolites and ultimately CO_2 and water. The test substance is removed from solution unless a stable and soluble metabolite is formed. There are several different instruments on the market for measurement of DOC [31] which make this a relatively easy method to use.

Determination of the carbon dioxide released from complete mineralization of the test substance is often used because it is general, and a specific analysis does not need to be developed. Another advantage of CO_2 determination is that it shows no intermediate metabolites are accumulating. Some studies have used other methods for determination of the extent of biodegradation. Two examples are foam formation in the biodegradation mixture and extraction of the unreacted surfactant with solvent and gravimetric determination of the surfactant. Thin layer chromatography has also been used as a way to follow the degradation products [29]. Another way of describing the biodegradation is to report the increase in biomass of the organisms which are utilizing the test substance for growth. The analytical method used is given for each study, along with their conclusions.

B. Test Systems

Many different types of test systems have been used for determination of the biodegradability of the polyoxyalkylene products. One of the most common is the shaken flask which contains the test substance and an inoculum from a wastewater treatment plant along with mineral nutrients. Aeration is provided by the shaking. The released carbon dioxide can be trapped and determined or the supernatant solution can be analyzed as described above. The Modified OECD Screening Test method, OECD 301E [30], is a commonly reported test procedure which uses a concentration of test substance chosen to give an initial DOC concentration of about 20 mg/L along with a rather small population of microorganisms (< 30 mg/L solids).

A test designed to simulate a wastewater treatment plant is the semicontinuous activated sludge system which has been modified and adopted by the OECD as method 302A [30]. This test is basically the procedure developed by the Soap and Detergent Association in the early 1960s [32]. This method can be used to expose microorganisms to the test substance for a long period of time as the units are operated on a daily fill-and-draw basis, fed daily with either synthetic sewage or settled municipal sewage and thus is useful for providing opportunity for acclimation of the microorganisms. The coupled units test, adopted by the OECD as method 303A [30], operates continuously to simulate operation of a wastewater treatment plant. It employs a relatively short hydraulic residence time of 3 or 6 hours and a flow of synthetic sewage with the test substance added. Other studies have used various enrichment procedures to isolate microorganisms which can utilize the test substance for growth.

C. Substances Tested

Pawlaczyk-Szpilowa et al. [33] studied the biodegradation of a variety of polyoxyalkylene products, the same selection as Karpinska-Smulikowska [24] (Table 2). All the compounds had a center POP segment with end POE segments. The biodegradation studies were carried out with 10,000 mg/L of test substance with an adapted inoculum from a river and aerated with sterile, carbon dioxide free air in a mineral medium. The high concentration of test substance was required because they extracted the mixture with chloroform and weighed the extracted material after evaporation to determine the extent of removal of polyoxyalkylene. The turbidity of the culture was also used to monitor bacterial growth. Some increase in turbidity of the culture was observed for all the samples. After 24 days incubation there was an apparent removal of test substance of up to 18%, based on the weight of material extracted. There was no apparent increase of carbonyl groups in the extracted material, as shown by infrared spectroscopy, and no change in the EO/PO ratio in the extracted material for most of the tested products, as observed determined by nuclear magnetic resonance. One product showed an increasing EO/PO ratio with incubation time. It is not clear how to interpret this, since the EO portion of the molecule is most available for biotransformation. Their conclusion was that since the increase in turbidity (microorganism growth) paralleled the decrease in extracted weight, the decrease was due to biodegradation. The observation that the apparent content of carbonyl decreased with time, was interpreted as indicating that oxidized compounds were more readily biodegraded than the parent compounds. This is consistent with the work described later in this review.

Kawai et al. [11], in their study of the biodegradation of POP examined bacterial growth on a variety of Epan products (POP with POE on each end of the molecule), using their isolate, *Corynebacterium* number 7. They observed the turbidity of cell suspensions to determine the extent of growth. Significant growth was only observed on those copolymers that contained a low percentage, about 10%, of POE. Products containing 50% or more of EO showed the same or less growth than the blank. Even the low fractions of EO apparently retarded growth as there was less growth than observed on any of the diol- or triol-type POP products tested.

Schoeberl et al. [13] summarized some otherwise unpublished environmental data on polyoxyalkylene surfactants. A product containing 20% EO and 80% PO (2500 mw), showed 1–10% of theoretical biochemical oxygen demand (BOD) in a closed bottle test. In the modified OECD screening test, this same product showed an 18% reduction in DOC. In the coupled units test, designed to simulate the performance of a wastewater treatment plant, there was only a 2±4% reduction in chemical oxygen demand (COD) for this product. This material also showed a 7% reduction in BIAS in the confirmatory test but a 32% reduction in BIAS in the screening test. These mixed results illustrate the difficulty in describing the partial biodegradation of a material using only nonspecific parameters.

Other products mentioned by Schoeberl et al. [13] included a 10% EO, 90% PO (2000 mw) product, which showed a 5% reduction in BIAS and a 15% reduction in COD in a Zahn-Wellens test. A 40% EO, 60% PO (4500 mw) product showed a 24% reduction in BIAS and an 11% reduction in DOC in the coupled units test. An 80% EO, 20% PO (9000 mw) product showed a 58% reduction in BIAS and 66% reduction in DOC in the coupled units test. In this series of tests, the extent of biodegradation increased with increasing POE content, which also included increased molecular weight.

Gerike and Fischer [34] included a polyoxyalkylene product as one of the test substances in their program for comparing biodegradation test methods. There is no description of the test substance and it was one of the compounds they listed as not biodegradable in all the tests. The results were as follows: coupled units test, 2±4% removal of DOC; Zahn-Wellens test, 19% removal of DOC in 14 days; Japanese Ministry for International Trade and Industry (MITI) test, 8% removal of DOC and 2% of theoretical biochemical oxygen demand in 14 days; Association Francoise de Normalization (AFNOR) test, 38% removal of DOC; Sturm test, no detected carbon dioxide released or DOC removed; OECD screening test, 11% removal of DOC; closed bottle test, 4% of theoretical oxygen consumed, BOD, in 30 days. The coupled units test result may be the same results as reported by Schoeberl et al. [13] and by Berth et al. [35]. Gerike and Fischer [36] report some additional work on the comparison of biodegradation test methods and report a 56% removal of DOC from an EO/PO polyoxyalkylene product in an "U.S. EPA activated sludge test." This test is basically the procedure developed by the Soap and Detergent Association in the early 1960s [32]. The test is designed to simulate a wastewater treatment plant with approximately one week allowed for adaptation to the test chemical. The mixed results for different test procedures show that the polyoxyalkylene is not readily biodegradable but is inherently biodegradable under the right conditions.

Pazdzioch et al. [37] reported that a surfactant composed of 14 BO and 23 EO units was biodegradable with 91.6% and 82.9% removed from a 20 mg/L solution by activated sludge. They commented that a relatively long period was required for adaptation and gave no more experimental details.

Sawyer et al. [38] and Bogan and Sawyer [39], in early studies on Pluronic F-68 (approximately 80% EO, 20% PO, 7850 mw) report very little oxygen demand in a five-day BOD test: 6% and 1.1% of theoretical oxygen demand with sewage seed and acclimated seed, respectively. This is probably close to the detection limit so that it is not clear if any biodegradation was observed. In a Warburg oxygen uptake test with 4600 mg/L activated sludge and 50 and 100 mg/L of Pluronic F-68, they report 3.6% of theoretical oxygen demand in 6 h.

Patterson et al. [29] report the biodegradation of a polyoxyalkylene EO/PO product (2000 mw), for comparison with a variety of polyglycols and alkyl alcohol ethoxylates. They used thin layer chromatography with visualization by a modified Dragendorff reagent and comparison of the spots from standard solutions. Disappearance of the copolymer was a little slower than that of POE (3300 mw) and much slower than lower molecular weight POE materials.

The removal of an unidentified olefin oxide surfactant, pictured as a polyoxyalkylene copolymer, by activated sludge was reported by Matsui et al. [40]. They used a typical activated sludge, mixed-liquor volatile suspended solids, concentration of 2200 mg/L. The reported results are 38, 37, and 55% removal of COD (by $KMnO_4$). The removal of DOC reported was 39, 45, and 45% after 2, 4, and 24 hours exposure, respectively. They also report only a few percent removal of POE in a similar experiment.

Brueschweiler [41] reported the biodegradation of a polyoxyalkylene block copolymer in the OECD screening test. The product tested had a molecular weight of 1750 and contained 10 oxyethylene units. After 8 days, approximately 10% reduction in surface activity and 15% reduction in BIAS was observed.

A polyoxyalkylene block copolymer described only as having a cloud point of 32 °C in a 1% solution was used as a negative control in a study of fatty alcohol polyoxyalkylene copolymers [42]; there was a 20% biodegradation in the screening test, as compared to over 90% for most fatty alcohol based products.

No information was found on the anaerobic biodegradation of polyoxyalkylene copolymers. However, the anaerobic biodegradation of POE homopolymers after up to 30 weeks acclimation has been reported [43,44]. The anaerobic biodegradation of POP homopolymers has not been reported. One could speculate that the polyoxyalkylene copolymers probably would biodegrade partially under anaerobic conditions; however, these experiments have not been reported to date.

D. Mechanisms

For information on the likely biochemistry involved in the aerobic biodegradation of the polyoxyalkylene copolymers, the only information is from analogies with the homopolyoxyalkylene products. Kawai [10] reviewed a series of studies on the biochemical reactions of both the POE and POP products. Apparently, some acclimation is required for biodegradation of polyoxyalkylene products [5,6,10]. As noted previously, POE compounds with molecular weights up to 20,000 can be biodegraded. POP compounds with molecular weights up to 4000 were reported to be biodegraded after extensive acclimation [11] in spite of their low water solubility. Kawai has shown that for the POP compounds, using dipropylene glycol as a model compound, the degradation apparently starts at the end of the chain with oxidation of the terminal alcohol to a ketone, rather than a hydrolytic cleavage of the ether link. The *Corynebacterium* sp. which is able to use POP, dipropylene glycol, and propylene glycol as sole carbon sources, was unable to utilize POE, showing the specificity of the enzymes involved. Similarly, bacteria able to utilize POE as a sole carbon source, are unable to utilize POP.

Clues to the biodegradation mechanism of these polymers can be obtained from the work with surfactants by Naylor et al. [45]. They worked with materials containing various amounts of PO along with the EO in fatty alcohol ethoxylates. The addition of only three or four PO (about 15% PO) units apparently adversely affected the biodegradation of the surfactants. The specificity of the enzyme systems for biodegradation of POE or POP are shown again in the work by Christopher et al. [46] on the enhancement of polyether biodegradation following exposure to unsaturated fatty acid diesters of the polyethers. The rate for biodegradation, CO_2 evolution, in diluted activated sludge increases over ten fold for POE (1000 mw), pre-exposed to diesterified POE 400 for 10 days. POP 1000 biodegradation is similarly enhanced by preexposure to the corresponding unsaturated diester. Diesters made from a saturated fatty acid seemed to have a much smaller effect. Using ^{14}C-labeled POE, Christopher et al. [46] showed that the majority of the radioactivity was released as CO_2 or remained in solution with only about 5% associated with the solids.

Apparently the major route of biodegradation of POE products is by oxidation of the terminal oxyalkylene –OH to an aldehyde by a dehydrogenase [10] followed by oxidation to a carboxylic acid and cleavage of the ether link. Steber and Wierich [47] studied the biodegradation of fatty alcohol ethoxylates and found primarily neutral and carboxylated POE metabolites. The identification of several C_2 units accounted for the stepwise depolymerization of POE. It was concluded that both hydrolytic and oxidative cleavage of the C_2 units was involved (Fig. 1).

Kawai et al. [48] observed the biodegradation of the different isomers of dipropylene glycol by a *Corynebacterium* sp. They found 1,2–propylene glycol

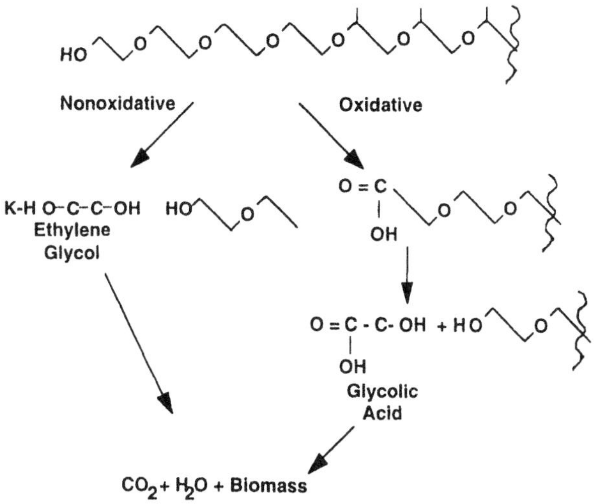

FIG. 1 Biodegradation pathways of polyoxyalkylene block copolymers. *Source*: adapted from [10] and [47].

as well as acetone derivatives as intermediate metabolites. Their study of the rates of degradation of the dipropylene glycol isomers showed that the secondary alcohol is oxidized most rapidly. Cell extracts from the culture grown in dipropylene glycol were able to oxidize not only the glycol but also a POP (2000 mw). From this they concluded that the degradation is primarily oxidative rather than hydrolytic.

In conclusion, the environmental behavior of the polyoxyalkylene block copolymers has been reviewed and they are found to resemble the homopolymers. They are reported to have low toxicity to aquatic organisms. The biodegradation data are mixed and suggest the copolymers are biodegradable, with acclimation sometimes important. These products are expected to remain in the aquatic environment with negligible volatilization into the atmosphere. Some adsorption on sediments is also expected which will reduce their mobility in the environment.

REFERENCES

1. A. L. Bridie, C. J. M. Wolff, and M. Winter, *Water Res. 13*:627 (1979).
2. H. Huekelekian and M. C. Rand, *Sewage Ind. Wastes 27*:1040 (1955).

3. P. Pitter, in *Chim. Phys. Appl. Prat. Ag. Surface, C. R. Congr. Int Deterg.*, 5th, Meeting September 1968, Vol. 1, pp. 115–123. Ediciones Unidas, S.A., Barcelona, Spain. Cited in *Chem. Abs. 74*:4917e.
4. R. J. Larson, R. T. Williams, and G. Swift, *Polym. Mater. Sci. Eng. 67*:348 (1992).
5. D. P. Cox, *Adv. Appl. Microbiol. 28*:173 (1978).
6. D. P. Cox and R. A. Conway, in *Proceedings of the Third International Biodegradation Symposium* (J. M. Sharpley and A. M. Kaplan, Eds.) Applied Science Pubs., London, 1976, pp. 835–841.
7. G. K. Watson and N. Jones, *Water Res. 11*:95 (1977).
8. P. Pitter, *Collect. Czech. Chem. Commun. 38*:2665 (1973). Cited in *Chem. Abs. 80*:4024k.
9. J. R. Haines and M. Alexander, *Appl. Microbiol. 29*:621 (1975).
10. F. Kawai, *CRC Crit. Rev. Biotech. 6*:273 (1987).
11. F. Kawai, K. Hanada, Y. Tani, and K. Ogata, *J. Ferment. Technol. 55*:89–96 (1977).
12. The Dow Chemical Company, unpublished data.
13. P. Schoeberl, K. J. Bock, M. Huber, and L. Huber, *Tenside Surf. Deterg. 25*:86 (1987).
14. S. W. Karickhoff, D. S. Brown, and T. A. Scott, *Water Res. 13*:241 (1979).
15. R. T. Podoll, K. C. Irwin, and S. Brendlinger, *Environ. Sci. Technol. 21*:562 (1987).
16. G. J. Fleer and J. Lyklema, in *Adsorption from Solution at the Solid/Liquid Interface* (G. D. Parfitt and C. H. Rochester, Eds.), Academic Press, London, 1983, pp. 153–220.
17. Medicinal Chemistry Project, Medchem Release 3.54, Daylight CIS, 1989.
18. A. Opperhuizen, E. W. v. d. Velde, F. A. P. C. Gobas, D. A. K. Liem, and J. M. D. v. d. Steen, *Chemosphere 14*:1871 (1985).
19. D. Mackay, W. Y. Shiu, and K. C. Ma, *Illustrated Handbook of Physical-Chemical Properties and Environmental Fate for Organic Chemicals*, Lewis Publishers, Chelsea, MI, 1992.
20. R. Atkinson, *Environ. Toxicol. Chem. 7*:435 (1988).
21. A. P. Altshuller, *J. Air Pollut. Contr. Assoc. 39*:704–708 (1989).
22. Uniform System for the Evaluation of Substances, Ministry of Housing, Spatial Planning and the Environment, Risk Assessment and Environmental Quality Division, The Hague, the Netherlands (1994).
23. E. Zucker, Standard Evaluation Procedure, Acute Toxicity Test for Freshwater Fish, EPA-540/9–85–006, June 1985 and Standard Evaluation Procedure, Acute Toxicity Test for Freshwater Invertebrates, EPA-540/9–85–005.
24. J. Karpinska-Smulikowska, *Tenside Surf. Deterg. 21*:243 (1984).

25. R. D. Swisher, *Surfactant Biodegradation*, Second Edition, Marcel Dekker, New York, 1987.
26. K. Burger, *Z. Anal. Chem. 196*:251 (1963).
27. R. Wickbold, *Tenside Surf. Deterg. 9*:173 (1972).
28. R. Wickbold, *Tenside Surf. Deterg. 10*:179 (1973).
29. S. J. Patterson, C. C. Scott, and K. B. E. Tucker, *J. Am. Oil Chem. Soc. 44*:407 (1967).
30. Organization for Economic Cooperation and Development, *OECD Guidelines for Testing of Chemicals*, OECD, Paris, 1993.
31. There are several different manufacturers of instruments for determination of total carbon in solution, including Shimadzu (Columbia, MD) and OI Analytical (College Station, TX).
32. C. M. Snow, *J. Am. Oil Chem. Soc. 42*:986 (1965).
33. M. Pawlaczyk-Szpilowa, J. Plucinski, J. Karpinska-Smulikowska, O. Staroojciec, and R. Janik, *Wiss. Z. Techn. Univers. Dresden 26*:1178 (1977).
34. P. Gerike and W. K. Fischer, *Ecotoxicol. Environ. Safety 3*:159 (1979).
35. P. Berth, P. Gerike, P. Gode and J. Steber, *Tenside Surf. Deterg. 25*:108 (1988).
36. P. Gerike and W. K. Fischer, *Ecotoxicol. Environ. Safety 5*:45 (1981).
37. W. Pazdzioch, J. Myszkowski, K. Szafraniak, and W. Goc, *Przemysl Chem. 60*:402 (1981).
38. C. N. Sawyer, R. H. Bogan, and J. R. Simpson, Ind. Engin. Chem. 48:236 (1956).
39. R. H. Bogan and C. N. Sawyer, *Sewage and Ind. Wastes 26*:1069 (1954).
40. S. Matsui, T. Murakami, T. Sasaki, Y. Hirose, and Y. Iguma, *Prog. Water Technol. 7*:645 (1975).
41. H. Brueschweiler, *Chimia 29*:31 (1975).
42. D. P. Karsa, J. Adamson, and R. P. Hadfield, *Chimicaoggi 10*:39 (1992).
43. D. F. Dwyer and J. M. Tiedje, *Appl. Environ. Microbiol. 46*:185 (1983).
44. B. Schink and M. Stieb, *Appl. Environ. Microbiol. 45*:1905 (1983).
45. C. G. Naylor, F. J. Castaldi, and B. J. Hayes, *J. Am. Oil Chem. Soc. 65*:1669 (1988).
46. L. J. Christopher, G. Holzer, and J. S. Hubbard, *Environ. Technol. 13*:521 (1992).
47. J. Steber and P. Wierich, *Appl. Environ. Microbiol. 49*:530 (1985).
48. F. Kawai, T. Okamoto, and T. Suzuki, *J. Ferment. Technol. 63*:239–244 (1985).

Index

Acclimation, bacterial, 244
Adsorption, in soil, 245
Aggregation (*see* Association)
Agricultural applications, 195
Alcohol ethoxylates (*see* Polarity index of alcohol ethoxylates)
Alcohols, as initiators, 7–9
Aldehydes, analysis with DNPH derivatives, 56
Aldol condensation, 22
Alkoxide ion:
  fraction of, 5
  nature of, 5
  reactivity of, 5–6
Alkylene oxide polymerization (*see* Polymerization of alkylene oxides)
Alkylene oxides, 2
Allyl alcohol, as isomerization product of PO, 19–20 (*see also* Analysis of unsaturates)
Amines, as initiators (*see* Initiator types, amines)
Analysis of unsaturates, 52–53
Antifoam agents (*see* Foam control agents)
Antioxidants, analysis of, 56–58
Applications of POA block copolymers, 186–202
Aquatic toxicity (*see* Toxicity, aquatic)
Area per molecule at the interface, 102–103
  effect of number of EO units, 109
  effect of POP block length, 109
  effect of temperature, 109
  of POP/POE versus POB/POE block copolymers, 177–178
Association:
  behavior of PO/EO block copolymers, dilute region, 75–111
  number, for Pluronic polyols, 89–93
Average EO block size, 16
Average molecular weight:
  by hydroxyl number, 49
  by size exclusion chromatography, 49–50

Bacterial abhesives, 190
Bacterial acclimation (*see* Acclimation, bacterial)
Base catalyst (*see* Catalyst, basic)
Bioaccumulation, 246
Biodegradation, of POA block copolymers, 249–255
  analytical methods, 249–250
  chemical mechanism, 254–255
  test protocols, 250–251
Block copolymers:
  preparation of, 3
  solution characterization methods, 72–73
Block length: (*see also* Block structure)
  of polyoxyethylene, 16
    effects on CMC, 81
    effects on melting point, 16
  of polyoxypropylene:
    effects on CMC, 80–81
Block structure, relation to oxide addition, 9–11
Block transition zone, effects of digest time, 21–22
Blood substitutes, 227
Burn wound dressings, 191
Butylene oxide (BO) (*see* Alkylene oxides)

C_{20} (*see* pC_{20})
Capping efficiency, 16
  with EO, 15–16
Carbonyl content:
  aldehydes, 56
  analysis by vibrational spectroscopy, 56
  in block copolymers, 56
Carboxylic acids (*see* Initiator types, carboxylic acids)
Catalyst:
  basic, 2
  choice of, 11–12
[Catalyst]
  coordination, 2
  Lewis acid, 2
  metal porphyrin, 2
Cement (*see* Petroleum applications, cement)
Chain propagation, 5
Chain transfer, acid-base, 5
Cleaners (*see* Detergent applications)
Cloud point, effects of salts, 93
CMC (*see* Critical micellization concentration)
CMC/C_{20}, 104–105, 110, 177–178
CMT (*see* Critical micelle temperature)
Coal applications (*see* Petroleum applications, coal dewatering)
Color, unwanted, 22
Copolymers, random, 3 (*see* also Block copolymers)
Corrosion, prevention of, 201
Cosmetics (*see* personal care applications)
Critical micelle temperature (CMT), 75–82
Critical micellization concentration, 75–82
  determination by dye solubilization, 110, (*see also* Dye solubilization technique)
  effects of impurities, 110
  effects of supplier on, 76–77
  effects of temperature on, 76
  of E-P-E block copolymers, 78–80
  surface tension at, 103–105
Crotyl alcohol, as rearrangement product of BO, 19–20
Crude oil demulsification (*see* Petroleum applications, demulsification)
Crystallization:
  of EPE, PEP, and PE block copolymers, 150–152

[Crystallization]
of POA block copolymers by DSC, 153–154
of POA block copolymers, modification of, 153
of polyoxyethylene homopolymers, 148–150
Crystals, liquid (*see* Liquid crystals)

Defoamers (*see* Foam control agents)
Dentifrice, 199
Detergent applications, 198–199
Digestion, of oxide, 21–22
Diols, 9–11
Dispersity (*see* Polydispersity)
Distribution, environmental, 246–247
Draize test, 220
Draves test (*see* Wetting)
Drug delivery, 186–188
Drug release, controlled, 188–190
Dye solubilization technique, 74

Effectiveness (*see* Critical micellization concentration, surface tension at)
Efficiency (*see*, pC_{20})
Emulsion polymerization (*see* Polymerization, emulsion)
Emulsions:
medical, 191–192
of perfluoro–surfactants, 192
of super-heavy oils, 194
End group (*see* Hydroxyl end group)
Enthalpy (*see* Micellization, enthalpy of)
Entropy (*see* Micellization, entropy of)
Environmental activity, of POA block copolymers, 244–255
Enzyme stabilization, 193
Equivalent weight (*see* Hydroxyl equivalent weight)
Ethylene oxide (EO) (*see* alkylene oxides)
Ethylene oxide/propylene oxide ratio:
by acetyl chloride treatment, 52
by hydrobromic acid treatment, 52
by vibrational spectroscopy, 39, 51
Ethyloxirane (*see* Alkylene oxides)
Extraneous initiation, 4, 12
effects of counterion, 19–20
effects on structure, 20–21
sources of, 16–20

Finishing (*see* Neutralization)
Flory's assumptions, 14–16
Fluorescence probes, 74
Foam control agents, 201
Foaming properties:
of Pluronic and Tetronic block copolymers, 171, 173
of POB/POE block copolymers, 172–173
Foams, polyurethane, 194–195
Fourier transform infrared spectroscopy (FTIR) (*see* Vibrational spectroscopy)
Free energy (*see* Micellization, free energy of)

Gas chromatography: (*see also* Inverse gas chromatography)
of chemical cleavage products, 52
of pyrolysis products, 52
Gel formation:
effects of additives on, 131–133
of POB/POE block copolymers, 133–135
of POP/POE block copolymers, 126–133
thermoreversible, 126–128, 188
Gel permeation chromatography (*see* Size exclusion chromatography)

Gels:
  applications of, 186–188
  for burn wound dressings, 191
Genotoxicity, 223–224
Gibbs adsorption equation, 102

HIV inhibition, 192
HLB (*see* Hydrophile-lipophile balance) 157
Hoffman elimination, 7,8
Hydration of POP block, effects of micelle formation, 77, 80
Hydrophile-lipophile balance (HLB), 157
  calculation of, 157, 214
  experimental determination of, 157
  limitation of, 157, 159
Hydroxyl end group, 15–16 (*see also* Hydroxyl number, by vibrational spectroscopy)
  controlling number of, 23–24
  conversion to ester, 24
  conversion to ether, 24
  conversion to tosylate, 24
  conversion to urethane, 24
  modification of, 22–25
  primary, 3
  primary-to-secondary ratio, 36
  secondary, 3
Hydroxyl equivalent weight, 9
  measurement of, 17
  target, 13
Hydroxyl number, by vibrational spectroscopy, 39, 42

Infrared spectroscopy (IR) (*see* Vibrational spectroscopy)
Initiation, extraneous (*see* Extraneous initiation)
Initiator:
  functionality, 9–11
  hydrophobicity, 9
[Initiator]
  impurities as, 16–18
  selection of, 9–11
  size, 9
  types:
    amines, 7–8
    carboxylic acids, 8
    noncatalyzed, 7
    phenolics, 8, 11–14
Interfacial tension (*see* Surface activity)
Interferometers, in Raman detectors, 39
Inverse gas chromatography, 157–158
Iodophors, 198
Isomerization: (*see also*, Extraneous initiation)
  effects of base, 19–20
  effects of temperature, 19
  of PO and BO, 19–20

Latex, stabilization of, 196
Light scattering:
  for obtaining association number, 89, 92
  techniques, 71–75
Liquid chromatography, 33, 55–56, 58
Liquid crystals, 112–118

Mammalian toxicity (*see* Toxicity, mammalian)
Mass spectrometry, 43–49
  by chemical ionization, 43
  by electron ionization, 43
  by electrospray ionization, 45–48
  of fragments from POP/POE copolymers, 44–45
  by ion cyclotron resonance, 46, 48, 49
  by matrix-assisted laser desorption ionization, 45, 48–49

[Mass spectrometry]
of pluronic polyols, by chemical ionization, 43–45
of pluronic polyols, by electrospray ionization, 46–48
by pyrolysis, 43
with size exclusion chromatography, 46
by time of flight, 48
Maximum surface pressure (*see* Critical micellization concentration, surface tension at)
Medical applications of POA block copolymers, 186–193
Melting behavior (*see* Crystallization)
Mercuric acetate, analysis of unsaturation, 53
Metabolic fate, mammalian, 223
Metal, cleaning of, 200–201
Methyloxirane (*see* Alkylene oxides)
Micelle formation (*see* Micellization)
Micelles:
association number, 89
EP compared to EPE copolymers, 102
formation, with POB/POE block copolymers, 133–135
hydrodynamic radius, 89
interaction, as function of temperature, 117, 119
parameters, methods of analysis, 74–75
shape of, in dilute solution, 93
shape transition, 93–95
structure, 89–93
Micellization:
of block copolymers, theoretical calculations, 135–136
effects of chain architecture, 98–102
effects of composition inhomogeneity, 95–98
enthalpy of, 82–87
[Micellization]
enthalpy of, by DSC, 87–88
entropy of, 83, 86–87
EPE compared to EBE copolymers, 177–178
free energy of, 86, 177–178
PEP (reverse) block copolymers, 99–102
thermodynamics of, 82–89
Modification:
of copolymers, 22–25
of polyether backbone, 25
Molecular structure, by NMR, 34–38
Molecular weight: (*see also* Average molecular weight)
distribution, 4–5
by field-flow fractionation, 51
limitations of, 4
number average, calculating, 11
Monols, 9–11
Mouthwash, 200
Mutagenicity, 223–224

Neutralization, 8–9
with acid, 9
over, 9, 22
under, 9, 22
Nonfouling surfaces, 190
Nonyl phenol ethoxylates (*see* Polarity index, of nonyl phenol ethoxylates)
Nuclear Magnetic resonance (NMR):
carbon, 36–38
determination of percent EO, 36
EO/PO distribution, 36, 51–52
Fourier transform, 36
of Pluronic Polyols, 36–37
proton, 35–36
sample purity, 35

Octanol/water partition coefficient, 245

Oil wells (*see* Petroleum applications)
Oncologic observations, 221–223, 227–228
Oral care (*see* Dentifrice)
Oxidation, during polymerization, 22
Oxirane (*see also* Alkylene oxides)
  ring opening, 3–4
Oxiranes, polymerization of (*see* Polymerization of alkylene oxides)
Oxygen carriers, fluorocarbon-based, 192, 227

Palatability, of POA block copolymers, 214
Paper: (*see also* Pulp)
  coating of, 196–197
  deinking of, 197
  sizing agents, 197
Partition coefficient (log Kow) (*see* Octanol/water partition coefficient)
pC_{20}, 103, 177–178
Peroxides, analysis of, 55
Personal care applications, 199–200
Petroleum applications, 193–194
  cement, 194
  coal dewatering, 193
  demulsification, 193
  oil recovery, enhanced, 194
Phase behavior, 75
  effects of block architecture, 122–126
  measured by small angle neutron scattering (SANS), 118–122
  of POP/POE block copolymers, 112–126
  in nonaqueous media, 126
  in relation to cloud point, 117
  of reverse and normal POP/POE block copolymers, 99–100
Phase diagrams: (*see also* Phase behavior)
  of Pluronic/water systems, 128–131
  of reverse POP/POE block copolymers, 124–125
Photography applications, 197
Physical properties, of polyoxyalkylene block copolymers, 147–163
Plastics applications, 194–195
Pluronic polyols:
  description of, 187
  nomenclature of, 152
  research activity on, 70
Poisson distribution, 6
  deviations from, 14
Polarity index, 157–161
  of alcohol ethoxylates, 158–160
  effects of hydroxyl end groups on, 159–161
  of nonyl phenol ethoxylates, 158–160
  of POA block copolymers, 156–161
  relation to calculated HLB, 159–160
Poloxamers, description of, 187
Polydispersity:
  effects on surfactancy, 16
  narrow, 4–5
  typical, 4
Polyethers, 2
  as chemical intermediates, 23
Polyethylene glycol (*see* Polyoxyethylene)
Polymer characterization, 12–13
Polymerization:
  of alkylene oxides, 3–4
  calculations, 11–12
  emulsion, 195–196
  of ethylene oxide in storage, 18
  planning, 11–12
  rate (*see* Reaction rate)

Polyoxybutylene (POB)
hydrophobicity of, 156
polarity of, 158–160
structure of, 148
viscosity of, 154–155
Polyoxyethylene (POE)
analysis of, in block copolymers, 58
crystallization (*see* Crystallization of polyoxyethylene homopolymers)
viscosity of, 154–155
Polyoxypropylene (POP):
block length, effects on CMC, 80–81
hydrophobe, compared to alkyl, 81
structure of, 148
viscosity of, 154–155
Potassium hydroxide, 8
Properties, of POB/POE versus POP/POE block copolymers, 173–178
Propylene oxide (PO) (*see* Alkylene oxides)
Pulp, synthetic, wetting of, 202

Qualitative analysis, 34

Raman spectroscopy, of Pluronic Polyols, 39–42
Reaction rate:
effects of alcohol acidity, 13–15
effects of sterics, 14–15
Rearrangement (*see* Isomerization)
Residual monomers/volatiles, analysis of, 53–55
by headspace GC, 54
by solid-phase extraction, 54
Ross-Miles foam test, 171
Rust inhibition (*see* Corrosion, prevention of)

Safety, in alkylene oxide polymerizations, 6–7
Salt:
effects on gel formation, 132
residual, 9
Shampoo, 200
Shock, hemorrhagic, 192
Size exclusion chromatography (SEC), 49–50
aggregation effects, 50
calibration, 50
detector types, 50
Sodium hydroxide, 8
Soil adsorption (*see* Adsorption, in soil)
Sol-gel transition, 120–121
Solubilization, 110–111
of hydrocarbons, 111
of polycyclic aromatic hydrocarbons, 202
Stability:
in acid-base media, 161–162
in oxidizing media, 162
thermal, 162–163
Stereoregularity, 2
Steric effects (*see* Reaction rate, effects of sterics)
Surface activity:
of POP/POE block copolymers, 102–110
of POP/POE versus POB/POE block copolymers, 175–178
table of values for POP/POE block copolymers, 104–105
Surface excess concentration, 102–105
Surface tension (*see* Surface activity)
Synthesis, nonideal, 13–22

Thermal stability (*see* Stability, thermal)
Thickeners, 201–202
Thrombosis, control of, 192

Toxicity, 212–239
aquatic, 247–249
gastrointestinal, 225–226
hematologic, 226–227
immunological, 224–225
mammalian:
dermal, 223, 236–239
inhalation, 215–218
intravenous, 218–219, 239
ocular, 219–221, 239
oral, 214–215, 221–223, 231–236
structure-activity relationships, 228–230
Transesterification, 8
Triols, 9–11
Tumors (*see* Oncologic observations)

Unsaturate formation (*see* Isomerization)
Unsaturates (see Analysis of unsaturates)

Vibrational spectroscopy, 38–43
attenuated total reflectance, 39
with chromatography, 42
of EO and PO functional groups, 43
[Vibrational spectroscopy]
Fourier transform, 38–43
of Pluronic polyols, 39–42
quantitative analysis, 42
Viscosity:
of block copolymers, 154–156, 213
of homopolymers, 154–155
Volatility, environmental, 246

Water:
in alkylene oxides, 18
analyzed by IR spectroscopy, 39, 42
effect on functionality, 18
effect on molecular weight, 17–18
as initiator, 17–18
solubility of polyoxyalkylene materials, 156, 213
Wetting:
Draves test, 164
of Pluronic block copolymers, 164–165
of POB/POE block copolymers, 167, 170–171
of reverse Pluronic and Tetronic copolymers, 166, 168–169
of Tetronic block copolymers, 164, 166

9 780824 797003